You Can Do Math:
Arithmetic Sequences

Copyright & Other Notices

Table of Contents

Introduction

For some time now, I have been tutoring both adults and children in math and science. As a result, I have discovered that many people of all ages have a latent interest (and talent) in mathematics that is somehow never got fully awoken while in school.

Because of my experience tutoring, I have been gradually developing a series of books on mathematical and science topics (a full list of my books can be found on my website at http://www.suniltanna.com). Most of my books are intended to teach specific topics and techniques, but I have also written others intended to awaken a student's interest in these subjects and broaden their horizons.

This book is about arithmetic sequences, which are ordered lists of numbers, where every successive number can be calculated from the previous number by adding the same value each time.

Before reading this book, I recommend you are comfortable with:

- Working with **fractions** - you may find my book **You Can Do Math:** Working with Fractions to be helpful in this regard.

- Working with **decimals**.

- Basic **algebra**.

- How to solve simple **simultaneous equations**.

- How to solve **quadratic equations**.

Let's get started…

Chapter 1: Introducing Arithmetic Sequences

In this chapter, we will introduce the idea of mathematical sequences, look at what makes a sequence an arithmetic sequence, and introduce some of the basic numerical properties of arithmetic sequences.

What is a Sequence?

In mathematics, a sequence is an ordered list of mathematical objects.

- The objects in the sequence can be numbers, shapes, vectors, equations, or anything else that mathematics deals with.
- The same object can appear at more than one position in the sequence (repetitions are allowed).
- Each object in the sequence is known as a term, element or member.
- The total number of terms in the sequence is known as the length of the sequence.
- Some sequences have a finite number of elements and are therefore of finite length. These are known as finite sequences.
- Some sequences go for ever and are therefore of infinite length. These are known as infinite sequences.

What is an Arithmetic Sequence?

In this book, we shall discuss arithmetic sequences (also known as arithmetic progressions), which are a very specific type of sequence:

- In an arithmetic sequence, the terms of the sequence are all numbers.
- Each successive term in an arithmetic sequence is calculated by adding the same value (known as the common difference) to the previous term. Note: the common difference is **not** necessarily whole number ("integer"), and it can be positive or negative.

Here are some examples of arithmetic sequences:

Example: 2, 5, 8, 11, 14, 17

This is an arithmetic sequence, because each successive term is calculated by adding 3 to the previous term. In this sequence, the common difference is 3.

Example: 2, 4, 6, 8, 10, 12, 14, 16, 18...

This is an arithmetic sequence, because each successive term is calculated by adding 2 to the previous term. In this sequence, the common difference is 2.

Example: 1.2, 1.6, 2, 2.4, 2.8, 3.2

This is an arithmetic sequence, because each successive term is calculated by adding 0.4 to the previous term. In this sequence, the common difference is 0.4

Example: 10, 6, 2, -2, -6, -10, -14

This is an arithmetic sequence, because each successive term is calculated by adding -4 to the previous term. In this sequence, the common difference is -4.

Example: 5.0, 4.3, 3.6, 2.9, 2.2...

This is an arithmetic sequence, because each successive term is calculated by adding -0.7 to the previous term. In this sequence, the common difference is -0.7.

Checking if a Sequence is an Arithmetic Sequence

You can whether any given sequence of numbers is an arithmetic sequence by calculating the difference between successive pair of terms in the sequence. If the difference is always the same for **every** pair of terms, then it is an arithmetic sequence. If the difference is **not** always the same, then it is **not** an arithmetic sequence (although it is still a sequence of some other kind).

Example: Is this an arithmetic sequence? 3, 8, 13, 18, 23

Solution:

- 8 - 3 = 5. There is a difference of 5 between the 2nd and 1st terms.
- 13 - 8 = 5. There is a difference of 5 between the 3rd and 2nd terms.
- 18 - 13 = 5. There is a difference of 5 between the 4th and 3rd terms.
- 23 - 18 = 5. There is a difference of 5 between the 5th and 4th terms.

- Since the difference between successive term is always 5, this is an arithmetic sequence with common difference of 5.

Example: Is this an arithmetic sequence? 1.9, 1.5, 1.1, 0.7, 0.3, -0.1, -0.5

Solution:

- 1.5 - 1.9 = -0.4. There is a difference of -0.4 between the 2nd and 1st terms.
- 1.1 - 1.5 = -0.4. There is a difference of -0.4 between the 3rd and 2nd terms.
- 0.7 - 1.1 = -0.4. There is a difference of -0.4 between the 4th and 3rd terms.
- 0.3 - 0.7 = -0.4. There is a difference of -0.4 between the 5th and 4th terms.
- -0.1 - 0.3 = -0.4. There is a difference of -0.4 between the 6th and 5th terms.
- -0.5 - -0.1 = -0.4. There is a difference of -0.4 between the 7th and 6th terms.
- Since the difference between successive term is always -0.4, this is an arithmetic sequence with common difference of -0.4.

Example: Is this an arithmetic sequence? 2, 4, 8, 16, 32

Solution:

- 4 - 2 = 2. There is a difference of 2 between the 2nd and 1st terms.
- 8 - 4 = 4. There is a difference of 4 between the 3rd and 2nd terms.
- At this point, we know the difference between each pair is **not** always the same, so this is **not** an arithmetic sequence.

Example: Is this an arithmetic sequence? 3, 5, 7, 11, 13, 17

Solution:

- 5 - 3 = 2. There is a difference of 2 between the 2nd and 1st terms.
- 7 - 5 = 2. There is a difference of 2 between the 3rd and 2nd terms.
- 11 - 7 = 4. There is a difference of 4 between the 4th and 3rd terms.
- At this point, we know the difference between each pair is **not** always the same, so this is **not** an arithmetic sequence.

Finding the Common Difference of an Arithmetic Sequence from Adjacent Terms

If you know any two adjacent terms in an arithmetic sequence you can find the common difference by subtracting the earlier term from the later term.

Example: In an arithmetic sequence, the 6th term is 19 and the 7th term is 25. What is the common difference?

Solution:

- Common difference = 25 - 19 = 6.

Example: In an arithmetic sequence, the 2nd term is 0.8 and the 3rd term is 0.1. What is the common difference?

Solution:

- Common difference = 0.1 - 0.8 = -0.7.

Finding the Common Difference of an Arithmetic Sequence from Non-Adjacent Terms

If you know any two non-adjacent terms in an arithmetic sequence you can find the common difference by subtracting the earlier term from the later term and then dividing by how many positions apart the two terms are.

Example: In an arithmetic sequence, the 2nd term is 11 and the 7th term is 26. What is the common difference?

Solution:

- The 7th and 2nd terms are 7 - 2 = 5 positions apart.
- Common difference = (26 - 11) ÷ 5 = 15 ÷ 5 = 3.

Example: In an arithmetic sequence, the 4th term is 2.1 and the 8th term is 0.9. What is the common difference?

Solution:

- The 8th and 4th terms are 8 - 4 = 4 positions apart.
- Common difference = (0.9 - 2.1) ÷ 4 = -1.2 ÷ 4 = -0.3.

Example: In an arithmetic sequence, the 3rd term is 36 and the 7th term is 21. What is the common difference?

Solution:

- The 7th and 3rd terms are 7 - 3 = 4 positions apart.
- Common difference = (21 - 36) ÷ 4 = -15 ÷ 4 = -3.75.

Finding the Next Terms in an Arithmetic Sequence

Providing you are given the common difference, or you are given enough information to calculate the common difference, you can calculate the next term in an arithmetic sequence. To find the next term, simply add the common difference to a known term. You can find the next term after that by adding the common difference again, and so on.

Example: An arithmetic sequence has a common difference of 9. The 5th term is 100. What are the 6th, 7th, 8th and 9th terms?

Solution:

- 6th term = 5th term + common difference = 100 + 9 = 109.
- 7th term = 6th term + common difference = 109 + 9 = 118.
- 8th term = 7th term + common difference = 118 + 9 = 127.
- 9th term = 8th term + common difference = 127 + 9 = 136.

Example: The 3rd and 4th terms of an arithmetic sequence are 7 and 11, what are the 5th and 6th terms?

Solution:

- Common difference = 11 - 7 = 4.
- 5th term = 4th term + common difference = 11 + 4 = 15.
- 6th term = 5th term + common difference = 15 + 4 = 19.

Example: The 4th and 7th terms of an arithmetic sequence are 25 and 1, what are the 8th and 9th terms?

Solution:

- The 7th and 4th terms are 7 - 4 = 3 positions apart.
- Common difference = (1 - 25) ÷ 3 = -24 ÷ 3 = -8.
- 8th term = 7th term + common difference = 1 + (-8) = -7.
- 9th term = 8th term + common difference = -7 + (-8) = -15.

Finding the Previous Terms in an Arithmetic Sequence

Providing you are given the common difference, or you are given enough information to calculate the common difference, you can calculate the previous term in an arithmetic sequence. To find the previous term, simply subtract the common difference from a known term. You can find the previous term before that by subtracting the common difference again, and so on.

Example: An arithmetic sequence has a common difference of 9. The 5th term is 100. What are the 4th, 3rd, 2nd and 1st terms?

Solution:

- 4th term = 5th term - common difference = 100 - 9 = 91.
- 3rd term = 4th term - common difference = 91 - 9 = 82.
- 2nd term = 3rd term - common difference = 82 - 9 = 73.
- 1st term = 2nd term - common difference = 73 - 9 = 64.

Example: The 3rd and 4th terms of an arithmetic sequence are 7 and 11, what are the 2nd and 1st terms?

Solution:

- Common difference = 11 - 7 = 4.
- 2nd term = 3rd term - common difference = 7 - 4 = 3.
- 1st term = 2nd term - common difference = 3 - 4 = -1.

Example: The 4th and 7th terms of an arithmetic sequence are 25 and 1, what are the 3rd, 2nd and 1st terms?

Solution:

- The 7th and 4th terms are 7 - 4 = 3 positions apart.
- Common difference = (1 - 25) ÷ 3 = -24 ÷ 3 = -8.
- 3rd term = 4th term - common difference = 25 - (-8) = 33.
- 2nd term = 3rd term - common difference = 33 - (-8) = 41.
- 1st term = 2nd term - common difference = 41 - (-8) = 49.

Finding Missing Terms in an Arithmetic Sequence

If you have some of the terms in an arithmetic sequence, but you are missing some other terms, you can find the missing terms by working forwards from a known term and adding the common difference each time, or by working backwards from a known term and subtracting the common difference each time.

A common mistake when finding missing terms in the middle of a sequence is to calculate the common difference incorrectly. For example, if shown part of an arithmetic sequence with two missing terms, a lot of people will **incorrectly** divide the difference between the surrounding terms by 2 instead of 3. This mistake is illustrated in our first example below:

Example: Find the missing terms in the following arithmetic sequence: 5 __ __ 23.

Common mistake:

- Many people will do 23 - 5 = 18, notice that there are two missing terms in the sequence, and therefore do 18 ÷ 2 = 9 to calculate the common difference. Then they will say that the 2nd term must be 5 + 9 = 14, and the 3rd term must be 14 + 9 = 23. **This is incorrect** because they have calculated the common difference incorrectly.

Solution:

- The 4th and 1st terms are 4 - 1 = 3 positions apart.
- Common difference = (23 - 5) ÷ 3 = 18 ÷ 3 = 6.
- 2nd term = 1st term + common difference = 5 + 6 = 11.
- 3rd term = 2nd term + common difference = 11 + 6 = 17.

Alternate solution:

- The 4th and 1st terms are 4 - 1 = 3 positions apart.
- Common difference = (23 - 5) ÷ 3 = 18 ÷ 3 = 6.
- 3rd term = 4th term - common difference = 23 - 6 = 17.
- 2nd term = 3rd term - common difference = 17 - 6 = 11.

Example: The 4th and 7th terms of an arithmetic sequence are 25 and 1, what are the 5th and 6th terms?

Solution:

- The 7th and 4th terms are 7 - 4 = 3 positions apart.
- Common difference = (1 - 25) ÷ 3 = -24 ÷ 3 = -8.
- 5th term = 4th term + common difference = 25 + (-8) = 17.
- 6th term = 5th term + common difference = 17 + (-8) = 9.

Alternate solution:

- The 7th and 4th terms are 7 - 4 = 3 positions apart.
- Common difference = (1 - 25) ÷ 3 = -24 ÷ 3 = -8.
- 6th term = 7th term - common difference = 1 - -8 = 9.

- 5th term = 6th term - common difference = 9 - -8 = 17.

Questions

Here are some questions about arithmetic sequences:

1. Which of the following from A to E are arithmetic sequences? (The correct answer may include none, one, or more than one of the options).

A. 1 2 4 8 16 32

B. 2 4 6 8 10 12 14 16

C. 1 1 2 3 5 8 13 21 34

D. 2 3 5 7 11 13 17 19 23 29

E. 1 4 9 16 25 36 49 64

2. Which of the following from A to E are arithmetic sequences? (The correct answer may include none, one, or more than one of the options).

A. 1 11 111 1111 11111 111111

B. 3.5 2.9 2.3 1.7 1.1

C. 0.1 0.2 0.4 0.4 0.6 0.8 0.9 1.0

D. 9 16 23 30 37 43

E. 3 5 9 17 33 65

3. Which of the following from A to E are arithmetic sequences? (The correct answer may include none, one, or more than one of the options).

A. 9 3 -3 -9 -15 -21

B. 2 2.5 3 3.6 4 4.7 5 5.8

C. 1980 1984 1988 1992 1996 2000 2004 2008 2012 2016

D. 1 2 3 4 5 6 7 8 9 10 11 12

E. 12 34 56 78 90

4. Which of the following from A to E are arithmetic sequences? (The correct answer may include none, one, or more than one of the options).

A. 9 7 5 3 1 -1 -3 -5 -7 -9

B. 47 35 23 11 -1 -13

C. 1 -1 2 -2 3 -3 4 -4 5 -5

D. 1 1 2 2 3 3 4 4 5 5 6 6

E. -1 0 1 2 3 4 5 6 7 8

5. Which of the following from A to E are arithmetic sequences? (The correct answer may include none, one, or more than one of the options).

A. 5 10 15 20 25 30 35 40 45 50

B. 5 15 20 25 35 40 45 50

C. 90 87 84 81 78 75 72 69

D. 0.3 0.7 1.1 1.5 1.9 2.3 2.7

E. -4 -15 -26 -37 -48 -59 -70 -81

6. What is the common difference in this arithmetic sequence: 3 2.3 1.6 0.9 0.2?

7. What is the common difference in this arithmetic sequence: 65 76 87 98 109?

8. The 2nd and 5th terms of an arithmetic sequence are 28 and 10. What is the common difference?

9. The 1st and 11th terms of an arithmetic sequence are 1 and 201. What is the common difference?

10. An arithmetic sequence begins 3 7 11. What are 4th, 5th and 6th terms?

11. An arithmetic sequence begins 19 12.5 6. What are the 4th, 5th and 6th terms?

12. The 2nd and 9th terms of an arithmetic sequence are 11 and 67. What are the 10th and 11th terms?

13. The 1st and 5th terms of an arithmetic sequence are 99 and 75. What are the 6th and 7th terms?

14. The 3rd and 5th terms of an arithmetic sequence are 77 and 66. What are the 1st and 2nd terms?

15. The 4th and 7th terms of an arithmetic sequence are 29 and 44. What are the 1st, 2nd and 3rd terms?

16. The 3rd and 8th terms of an arithmetic sequence are 3.4 and 4.9. What are the 1st and 2nd terms?

17. Fill in the missing numbers in the following arithmetic sequence: __ 23 __ __ __ 59.

18. Fill in the missing numbers in the following arithmetic sequence: __ __ __ 0.7 __ 0.1 __ __.

19. Fill in the missing numbers in the following arithmetic sequence: __ __ __ -0.3 __ 0.5 __ __.

20. Fill in the missing numbers in the following arithmetic sequence: __ __ __ -17 __ __ -29 __ __.

Answers to Chapter 1 Questions

1. B as this is the only option where the difference between successive terms is constant.

2. B and D, as these are the options where the difference between successive terms is constant.

3. A, C, D and E, as these are the only options where the difference between successive terms is constant.

4. A, B and E, as these are the only options where the difference between successive terms is constant.

5. A, C, D and E, as these are the only options where the difference between successive terms is constant.

6. Solution:

   - Common difference = 2.3 - 3 = -0.7.

7. Solution:

   - Common difference = 76 - 65 = 11.

8. Solution:

   - The 5th and 2nd terms are 5 - 2 = 3 positions apart.
   - Common difference = (10 - 28) ÷ 3 = -18 ÷ 3 = -6.

9. Solution:

   - The 11th and 1st terms are 11 - 1 = 10 positions apart.
   - Common difference = (201 - 1) ÷ 10 = 200 ÷ 10 = 20.

10. Solution:

- Common difference = 7 - 3 = 4.
- 4th term = 3rd term + common difference = 11 + 4 = 15.
- 5th term = 4th term + common difference = 15 + 4 = 19.
- 6th term = 5th term + common difference = 19 + 4 = 23.

11. Solution:

- Common difference = 12.5 - 19 = -6.5.
- 4th term = 3rd term + common difference = 6 + (-6.5) = -0.5.
- 5th term = 4th term + common difference = -0.5 + (-6.5) = -7.
- 6th term = 5th term + common difference = -7 + (-6.5) = -13.5.

12. Solution:

- The 9th and 2nd terms are 9 - 2 = 7 positions apart.
- Common difference = (67 - 11) ÷ 7 = 56 ÷ 7 = 8.
- 10th term = 9th term + common difference = 67 + 8 = 75.
- 11th term = 10th term + common difference = 75 + 8 = 83.

13. Solution:

- The 5th and 1st terms are 5 - 1 = 4 positions apart.
- Common difference = (75 - 99) ÷ 4 = -24 ÷ 4 = -6.
- 6th term = 5th term + common difference = 75 + (-6) = 69.
- 7th term = 6th term + common difference = 69 + (-6) = 63.

14. Solution:

- The 5th and 3rd terms are 5 - 3 = 2 positions apart.
- Common difference = (66 - 77) ÷ 2 = -11 ÷ 2 = -5.5.
- 2nd term = 3rd term - common difference = 77 - (-5.5) = 82.5.
- 1st term = 2nd term - common difference = 82.5 - (-5.5) = 88.

15. Solution:

- The 7th and 4th terms are 7 - 4 = 3 positions apart.

- Common difference = (44 - 29) ÷ 3 = 15 ÷ 3 = 5.
- 3rd term = 4th term - common difference = 29 - 5 = 24.
- 2nd term = 3rd term - common difference = 24 - 5 = 19.
- 1st term = 2nd term - common difference = 19 - 5 = 14.

16. Solution:

- The 8th and 3rd terms are 8 - 3 = 5 positions apart.
- Common difference = (4.9 - 3.4) ÷ 5 = 1.5 ÷ 5 = 0.3.
- 2nd term = 3rd term - common difference = 3.4 - 0.3 = 3.1.
- 1st term = 2nd term - common difference = 3.1 - 0.3 = 2.8.

17. Solution:

- The 6th and 2nd terms are 6 - 2 = 4 positions apart.
- Common difference = (59 - 23) ÷ 4 = 36 ÷ 4 = 9.
- 1st term = 2nd term - common difference = 23 - 9 = 14.
- 3rd term = 2nd term + common difference = 23 + 9 = 32.
- 4th term = 3rd term + common difference = 32 + 9 = 41.
- 5th term = 4th term + common difference = 41 + 9 = 50.
- Completed arithmetic sequence: <u>14</u> 23 <u>32</u> <u>41</u> <u>50</u> 59.

18. Solution:

- The 6th and 4th terms are 6 - 4 = 2 positions apart.
- Common difference = (0.1 - 0.7) ÷ 2 = -0.6 ÷ 2 = -0.3.
- 3rd term = 4th term - common difference = 0.7 - (-0.3) = 1.0.
- 2nd term = 3rd term - common difference = 1.0 - (-0.3) = 1.3.
- 1st term = 2nd term - common difference = 1.3 - (-0.3) = 1.6.
- 5th term = 4th term + common difference = 0.7 + (-0.3) = 0.4.
- 7th term = 6th term + common difference = 0.1 + (-0.3) = -0.2.
- 8th term = 7th term + common difference = -0.2 + (-0.3) = -0.5.
- Completed arithmetic sequence: <u>1.6</u> <u>1.3</u> <u>1.0</u> 0.7 <u>0.4</u> 0.1 <u>-0.2</u> <u>-0.5</u>.

19. Solution:

- The 6th and 4th terms are 6 - 4 = 2 positions apart.
- Common difference = (0.5 - -0.3) ÷ 2 = 0.8 ÷ 2 = 0.4.
- 3rd term = 4th term - common difference = -0.3 - 0.4 = -0.7.

- 2nd term = 3rd term - common difference = -0.7 - 0.4 = -1.1.
- 1st term = 2nd term - common difference = -1.1 - 0.4 = -1.5.
- 5th term = 4th term + common difference = -0.3 + 0.4 = 0.1.
- 7th term = 6th term + common difference = 0.5 + 0.4 = 0.9.
- 8th term = 7th term + common difference = 0.9 + 0.4 = 1.3.
- Completed arithmetic sequence: <u>-1.5</u> <u>-1.1</u> <u>-0.7</u> -0.3 <u>0.1</u> 0.5 <u>0.9</u> <u>1.3</u>.

20. Solution:

- The 7th and 4th terms are 7 - 4 = 3 positions apart.
- Common difference = (-29 - -17) ÷ 3 = -12 ÷ 3 = -4.
- 3rd term = 4th term - common difference = -17 - (-4) = -13.
- 2nd term = 3rd term - common difference = -13 - (-4) = -9.
- 1st term = 2nd term - common difference = -9 - (-4) = -5.
- 5th term = 4th term + common difference = -17 + (-4) = -21.
- 6th term = 5th term + common difference = -21 + (-4) = -25.
- 8th term = 7th term + common difference = -29 + (-4) = -33.
- 9th term = 8th term + common difference = -33 + (-4) = -37.
- Completed arithmetic sequence: <u>-5</u> <u>-9</u> <u>-13</u> -17 <u>-21</u> -25 -29 <u>-33</u> <u>-37</u>.

Chapter 2: Finding and Using the Formula for a Particular Arithmetic Sequence

In Chapter 1, we looked at what arithmetic sequences are, and how you can use various ad hoc methods to find next, previous or missing terms in a known sequence. In this chapter, we will look at how to develop and use a formula for calculating the terms of an arithmetic sequence.

Finding the Formula

Finding the formula for any arithmetic sequence is easy:

- **Step 1:** Find the common difference (using one of the methods discussed in Chapter 1).
- **Step 2:** Subtract the common difference from the 1st term, to find what would have been the zeroth term (the zeroth term is the term that would have preceded the 1st term if there had been such a term).
- **Step 3:** The formula for term k in the sequence is: term(k) = zeroth-term + common-difference × k.

Note: In this book we use k to indicate the position (starting from 1) of a particular term in the sequence. Some other books use n for this purpose. We have avoided using n, because n is generally used (including in this book) to indicate the total number of terms in a sequence.

Example: An arithmetic sequence begins 5, 13, 21. What is the formula for this sequence?

Solution:

- **Step 1:** Common difference = 13 - 5 = 8.
- **Step 2:** Zeroth term = 5 - 8 = -3.
- **Step 3:** Term(k) = zeroth-term + common-difference × k = -3 + 8k.

Example: The 2nd and 5th terms in an arithmetic sequence are 20 and 2. What is the formula for this sequence?

Solution:

- **Step 1:** Find the common difference...

- The 5th and 2nd terms are 5 - 2 = 3 positions apart.
- Common difference = (2 - 20) ÷ 3 = -18 ÷ 3 = -6.
- **Step 2:** Find the zeroth term...
- 1st term = 2nd term - common difference = 20 - (-6) = 26.
- Zeroth term = 1st term - common difference = 26 - (-6) = 32.
- **Step 3:** Term(k) = zeroth-term + common-difference × k = 32 - 6k.

Example: An arithmetic sequence (with some missing numbers) begins __ __ 14 __ __ 26. What is the formula for this sequence?

Solution:

- **Step 1:** Find the common difference...
- The 6th and 3rd terms are 6 - 3 = 3 positions apart.
- Common difference = (26 - 14) ÷ 3 = 12 ÷ 3 = 4.
- **Step 2:** Find the zeroth term...
- 2nd term = 3rd term - common difference = 14 - 4 = 10.
- 1st term = 2nd term - common difference = 10 - 4 = 6.
- Zeroth term = 1st term - common difference = 6 - 4 = 2.
- **Step 3:** Term(k) = zeroth-term + common-difference × k = 2 + 4k.

Using the Formula to Find a Particular Term

Once you have a formula for finding terms in an arithmetic sequence, you can use it to find the value of **any** term in that sequence: simply substitute in the term number for k.

Example: The value of the k'th term in an arithmetic sequence is given by Term(k) = 7 + 3k. What are the values of the 5th, 11th, 20th, 30th and 100th terms?

Solution:

- 5th term = Term(5) = 7 + 3 × 5 = 7 + 15 = 22.
- 11th term = Term(11) = 7 + 3 × 11 = 7 + 33 = 40.
- 20th term = Term(20) = 7 + 3 × 20 = 7 + 60 = 67.
- 30th term = Term(30) = 7 + 3 × 30 = 7 + 90 = 97.
- 100th term = Term(100) = 7 + 3 × 100 = 7 + 300 = 307.

Example: An arithmetic sequence begins 15, 8, 1. What are the values of the 5th, 11th and 100th terms?

Solution:

- Common difference = 8 - 15 = -7.
- Zeroth term = 1st term - common difference = 15 - (-7) = 22.
- Term(k) = zeroth-term + common-difference × k = 22 + (-7)k = 22 - 7k.
- 5th term = Term(5) = 22 - 7 × 5 = 22 - 35 = -13.
- 11th term = Term(11) = 22 - 7 × 11 = 22 - 77 = -55.
- 100th term = Term(100) = 22 - 7 × 100 = 22 - 700 = -678.

Using the Formula to Check Whether and Where a Particular Term is in a Sequence

Once we have the formula for calculating terms in an arithmetic sequence, we can use this formula to determine whether any particular number is a term in the sequence, and if so, its position in the sequence.

- **Step 1:** Equate the formula for term(k) to the number that you want to check.
- **Step 2:** Treating k as an unknown variable in the equation, solve for k (note: some basic algebra will be needed).
- **Step 3:** If k turns out to be a whole number ("integer"), this is the position of the number in the sequence. If k is **not** a whole number, then the number is **not** in the sequence.

Example: The value of the k'th term in an arithmetic sequence is given by Term(k) = 3 + 5k. Is 46 in the sequence, and if so, at what position?

Solution:

- **Step 1:** 3 + 5k = 46.
- **Step 2:** Solve the equation…
- 5k = 46 - 3.
- 5k = 43.
- k = 43 ÷ 5.
- k = 8.6.
- **Step 3:** Since k is **not** an integer, this means 46 is **not** in the sequence.

Example: An arithmetic sequence begins 19, 26, 33, 40. Is 159 in the sequence, and if so, at what position?

Solution:

- **First find the formula for the arithmetic sequence...**
- Common difference = 26 - 19 = 7.
- Zeroth term = 1st term - common difference = 19 - (7) = 12.
- Term(k) = zeroth-term + common-difference × k = 12 + 7k.
- **Step 1:** 12 + 7k = 159.
- **Step 2:** Solve the equation...
- 7k =159 - 12.
- 7k = 147.
- k = 147 ÷ 7.
- k = 21.
- **Step 3:** Since k is an integer, this means 159 does appear in the sequence at position 21.

Using the Formula to Find the First Term Greater Than a Target Value

In an arithmetic sequence with a positive common difference, each successive term in the sequence will have a higher value than the term that precedes it. A common requirement is to find the position of the first term in the sequence that equals or exceeds a particular target value. This type of question can be solved using a similar method to that used for locating the position of a term in the sequence:

- **Step 1:** Equate the formula for term(k) to the target value.
- **Step 2:** Treating k as an unknown variable in the equation, solve for k (note: some basic algebra will be needed).
- **Step 3:** If k is a whole number ("integer"), this is the position of the target value in the sequence. If k is **not** an integer, then rounding k down to the next lower integer will give the position of the last term in the sequence which is less than the target value and rounding k up to the next higher integer will give the position of the first term in the sequence which is greater than the target value.

Example: The value of the k'th term in an arithmetic sequence is given by Term(k) = 13 + 5k. What is the position and value of last term less than 100, and position and value of the first term over 100?

Solution:

- **Step 1:** $13 + 5k = 100$.
- **Step 2:** Solve the equation...
- $5k = 100 - 13$.
- $5k = 87$.
- $k = 87 \div 5$.
- $k = 17.4$.
- **Step 3:** The 17th term is therefore the last term less than 100, and the 18th term is the first term over 100.
- **Values of the Terms...**
- $\text{Term}(17) = 13 + 5 \times 17 = 13 + 85 = 98$.
- $\text{Term}(18) = 13 + 5 \times 18 = 13 + 90 = 103$.

Example: The 2nd and 6th terms of an arithmetic sequence are 23 and 51. What is the position and value of last term less than 1000, and position and value of the first term over 1000?

Solution:

- **First find the formula for the arithmetic sequence...**
- The 6th and 2nd terms are $6 - 2 = 4$ positions apart.
- Common difference $= (51 - 23) \div 4 = 28 \div 4 = 7$.
- 1st term = 2nd term - common difference $= 23 - 7 = 16$.
- Zeroth term = 1st term - common difference $= 16 - 7 = 9$.
- $\text{Term}(k)$ = zeroth-term + common-difference $\times k = 9 + 7k$.
- **Step 1:** $9 + 7k = 1000$
- **Step 2:** Solve the equation...
- $7k = 1000 - 9$.
- $7k = 991$.
- $k = 991 \div 7$.
- $k = 141.5714286...$
- **Step 3:** The 141st term is therefore the last term less than 1000, and the 142nd term is the first term over 1000.
- **Values of the Terms...**
- $\text{Term}(141) = 9 + 7 \times 141 = 9 + 987 = 996$.
- $\text{Term}(142) = 9 + 7 \times 142 = 9 + 994 = 1003$.

Using the Formula to Find the First Term Less Than a Target Value

In an arithmetic sequence with a negative common difference, each successive term in the sequence will have a lower value than the previous term. A common requirement is to find the position of the 1st term in the sequence that equals or is under a particular target value. This type of question can be solved using a similar method to that used for locating the position of a term in the sequence:

- **Step 1:** Equate the formula for term(k) to the target value.
- **Step 2:** Treating k as an unknown variable in the equation, solve for k (note: some basic <u>algebra</u> will be needed).
- **Step 3:** If k is a whole number ("integer"), this is the position of the target value in the sequence. If k is **not** an integer, then rounding k down to the next lower integer will give the position of the last term in the sequence which is greater than the target value and rounding k up to the next higher integer will give the position of the first term in the sequence which is less than the target value.

Example: The value of the k'th term in an arithmetic sequence is given by Term(k) = 73 - 5k. What is the position and value of last term greater than 0, and position and value of the first term below 0?

Solution:

- **Step 1:** 73 - 5k = 0.
- **Step 2:** Solve the equation...
- -5k = 0 - 73.
- -5k = -73.
- 5k = 73.
- k = 73 ÷ 5.
- k = 14.6.
- **Step 3:** The 14th term is therefore the last term greater than 0, and the 15th term is the first term below 0.
- **Values of the Terms...**
- Term(14) = 73 - 5 × 14 = 73 - 70 = 3.
- Term(15) = 73 - 5 × 15 = 73 - 75 = -2.

Example: The 2nd and 5th terms of an arithmetic sequence are 925 and 901. What is the position and value of last term greater than 100, and position and value of the first term below 100?

Solution:

- **First find the formula for the arithmetic sequence...**
- The 5th and 2nd terms are 5 - 2 = 3 positions apart.

- Common difference = (901 - 925) ÷ 3 = (-24) ÷ 3 = -8.
- 1st term = 2nd term - common difference = 925 - (-8) = 933.
- Zeroth term = 1st term - common difference = 933 - (-8) = 941.
- Term(k) = zeroth-term + common-difference × k = 941 - 8k.
- **Step 1:** 941 - 8k = 100
- **Step 2:** Solve the equation...
- -8k = 100 - 941.
- -8k = -841.
- 8k = 841.
- k = 841 ÷ 8.
- k = 105.125.
- **Step 3:** The 105th term is therefore the last term greater than 100, and the 106th term is the first term below 100.
- **Values of the Terms...**
- Term(105) = 941 - 8 × 105 = 941 - 840 = 101.
- Term(106) = 941 - 8 × 106 = 941 - 848 = 93.

Questions

Can you solve these questions using arithmetic sequence formulae?

1. Find the formula for the k'th term of the following arithmetic sequence: 12 7 2 -3. Using the formula what is the 10th term of this sequence?

2. Find the formula for the k'th term of the following arithmetic sequence: 0.9 1.5 2.1 2.7. Using the formula what is the 15th term of this sequence?

3. Find the formula for the k'th term of the following arithmetic sequence: 6 1 -4 -9. Using the formula what is the 11th term of this sequence?

4. Find the formula for the k'th term of the following arithmetic sequence: 16 10 4. Using the formula what is the 19th term of this sequence?

5. Here is an arithmetic sequence containing some missing terms: 9 __ __ 21. What is the formula for the k'th term of this following arithmetic sequence? Using the formula what is the 13th term of this sequence?

6. Find the formula for the k'th term of the following arithmetic sequence: 13 20 27 34. Is 105 a term in the sequence, and if so, at what position does it appear?

7. Find the formula for the k'th term of the following arithmetic sequence: 76 68 60 52. Is -28 a term in the sequence, and if so, at what position does it appear?

8. Find the formula for the k'th term of the following arithmetic sequence: 101 115 129 143. Is 533 a term in the sequence, and if so, at what position does it appear?

9. Here is an arithmetic sequence containing some missing terms: __ 10 __ 22. What is the formula for the k'th term of this following arithmetic sequence? Is 172 a term in the sequence, and if so, at what position does it appear?

10. Here is an arithmetic sequence containing some missing terms: __ __ 21 __ 29. What is the formula for the k'th term of this following arithmetic sequence? Is 149 a term in the sequence, and if so, at what position does it appear?

11. Find the formula for the k'th term of the following arithmetic sequence: 19 25 31 37. What is the position and value of last term less than 1000, and position and value of the first term over 1000?

12. Find the formula for the k'th term of the following arithmetic sequence: 211 208 205 202. What is the position and value of last term greater than 0, and position and value of the first term below 0?

13. Find the formula for the k'th term of the following arithmetic sequence: 2.9 3.6 4.3 5.0. What is the position and value of last term less than 100, and position and value of the first term over 100?

14. Here is an arithmetic sequence containing some missing terms: 11 __ __ 29. What is the formula for the k'th term of this following arithmetic sequence? What is the position and value of last term less than 1000, and position and value of the first term over 1000?

15. Here is an arithmetic sequence containing some missing terms: 147 __ __ __ 131. What is the formula for the k'th term of this following arithmetic sequence? What is the position and value of last term greater than 0, and position and value of the first term below 0?

16. Find the formula for the k'th term of the following arithmetic sequence: -67 -61 -55. What is the position and value of last term less than 100, and position and value of the first term over 100?

17. Find the formula for the k'th term of the following arithmetic sequence: -3 -11 -19 -27 -33. What is the position and value of last term greater than -200, and position and value of the first term below -200?

18. Find the formula for the k'th term of the following arithmetic sequence: 141 135 129 123 117. What is the position and value of last term greater than 0, and position and value of the first term below 0?

19. Here is an arithmetic sequence containing some missing terms: __ 23 __ __ 56. What is the formula for the k'th term of this following arithmetic sequence? What is the position and value of last term less than 1000, and position and value of the first term over 1000?

20. Here is an arithmetic sequence containing some missing terms: 112 104 96. What is the formula for the k'th term of this following arithmetic sequence? What is the position and value of last term greater than 0, and position and value of the first term below 0?

Answers to Chapter 2 Questions

1. Solution:

- Common difference = 7 - 12 = -5.
- Zeroth term = 1st term - common difference = 12 - (-5) = 17.
- Term(k) = zeroth-term + common-difference × k = 17 - 5k.
- Term(10) = 17 - 5 × 10 = 17 - 50 = -33.

2. Solution:

- Common difference = 1.5 - 0.9 = 0.6.
- Zeroth term = 1st term - common difference = 0.9 - 0.6 = 0.3.
- Term(k) = zeroth-term + common-difference × k = 0.3 + 0.6k.
- Term(15) = 0.3 + 0.6 × 15 = 0.3 + 9 = 9.3.

3. Solution:

- Common difference = 1 - 6 = -5.
- Zeroth term = 1st term - common difference = 6 - (-5) = 11.
- Term(k) = zeroth-term + common-difference × k = 11 - 5k.
- Term(11) = 11 - 5 × 11 = 11 - 55 = -44.

4. Solution:

- Common difference = 10 - 16 = -6.
- Zeroth term = 1st term - common difference = 16 - (-6) = 22.
- Term(k) = zeroth-term + common-difference × k = 22 - 6k.
- Term(19) = 22 - 6 × 19 = 22 - 114 = -92.

5. Solution:

- The 4th and 1st terms are 4 - 1 = 3 positions apart.
- Common difference = (21 - 9) ÷ 3 = 12 ÷ 3 = 4.
- Zeroth term = 1st term - common difference = 9 - 4 = 5.
- Term(k) = zeroth-term + common-difference × k = 5 + 4k.
- Term(13) = 5 + 4 × 13 = 5 + 52 = 57.

6. Solution:

- Common difference = 20 - 13 = 7.
- Zeroth term = 1st term - common difference = 13 - 7 = 6.
- Term(k) = zeroth-term + common-difference × k = 6 + 7k.
- **Step 1:** 6 + 7k = 105.
- **Step 2:** Solve the equation...
- 7k = 105 - 6.
- 7k = 99.
- k = 99 ÷ 7.
- k = 14.14285714...
- **Step 3:** Since k is **not** an integer, this means 105 is **not** in the sequence.

7. Solution:

- Common difference = 68 - 76 = -8.
- Zeroth term = 1st term - common difference = 76 - (-8) = 84.
- Term(k) = zeroth-term + common-difference × k = 84 - 8k.
- **Step 1:** 84 - 8k = -28.
- **Step 2:** Solve the equation...
- -8k = -28 - 84.
- -8k = -112.
- 8k = 112.
- k = 112 ÷ 8.
- k = 14.
- **Step 3:** Since k is an integer, this means -28 does appear in the sequence at position 14.

8. Solution:

- Common difference = 115 - 101 = 14.
- Zeroth term = 1st term - common difference = 101 - 14 = 87.
- Term(k) = zeroth-term + common-difference × k = 87 + 14k.
- **Step 1:** 87 + 14k = 533.
- **Step 2:** Solve the equation...
- 14k = 533 - 87.
- 14k = 446.
- k = 446 ÷ 14.
- k = 31.85714286...
- **Step 3:** Since k is **not** an integer, this means 533 is **not** in the sequence.

9. Solution:

- The 4th and 2nd terms are 4 - 2 = 2 positions apart.
- Common difference = (22 - 10) ÷ 2 = 12 ÷ 2 = 6.
- 1st term = 2nd term - common difference = 10 - 6 = 4.
- Zeroth term = 1st term - common difference = 4 - 6 = -2.
- Term(k) = zeroth-term + common-difference × k = -2 + 6k.
- **Step 1:** -2 + 6k = 172.
- **Step 2:** Solve the equation...
- 6k = 172 + 2.
- 6k = 174.
- k = 174 ÷ 6.
- k = 29.
- **Step 3:** Since k is an integer, this means 172 does appear in the sequence at position 29.

10. Solution:

- The 5th and 3rd terms are 5 - 3 = 2 positions apart.
- Common difference = (29 - 21) ÷ 2 = 8 ÷ 2 = 4.
- 2nd term = 3rd term - common difference = 21 - 4 = 17.
- 1st term = 2nd term - common difference = 17 - 4 = 13.
- Zeroth term = 1st term - common difference = 13 - 4 = 9.
- Term(k) = zeroth-term + common-difference × k = 9 + 4k.
- **Step 1:** 9 + 4k = 149.
- **Step 2:** Solve the equation...
- 4k = 149 - 9.
- 4k = 140.
- k = 140 ÷ 4.
- k = 35.
- **Step 3:** Since k is an integer, this means 149 does appear in the sequence at position 35.

11. Solution:

- Common difference = 25 - 19 = 6.
- Zeroth term = 1st term - common difference = 19 - 6 = 13.
- Term(k) = zeroth-term + common-difference × k = 13 + 6k.
- **Step 1:** 13 + 6k = 1000.
- **Step 2:** Solve the equation...
- 6k = 1000 - 13.
- 6k = 987.
- k = 987 ÷ 6.

- $k = 164.5$.
- **Step 3:** The 164th term is therefore the last term less than 1000, and the 165th term is the first term over 1000.
- **Values of the Terms...**
- Term(164) = 13 + 6 × 164 = 13 + 984 = 997.
- Term(165) = 13 + 6 × 165 = 13 + 990 = 1003.

12. Solution:

- Common difference = 208 - 211 = -3.
- Zeroth term = 1st term - common difference = 211 - (-3) = 214.
- Term(k) = zeroth-term + common-difference × k = 214 - 3k.
- **Step 1:** 214 - 3k = 0.
- **Step 2:** Solve the equation...
- -3k = 0 - 214.
- -3k = -214.
- 3k = 214.
- k = 214 ÷ 3.
- k = 71.33333333...
- **Step 3:** The 71st term is therefore the last term greater than 0, and the 72nd term is the first term below 100.
- **Values of the Terms...**
- Term(71) = 214 - 3 × 71 = 214 - 213 = 1.
- Term(72) = 214 - 3 × 72 = 214 - 216 = -2.

13. Solution:

- Common difference = 3.6 - 2.9 = 0.7.
- Zeroth term = 1st term - common difference = 2.9 - 0.7 = 2.2.
- Term(k) = zeroth-term + common-difference × k = 2.2 + 0.7k.
- **Step 1:** 2.2 + 0.7k = 100.
- **Step 2:** Solve the equation...
- 0.7k = 100 - 2.2.
- 0.7k = 97.8.
- k = 97.8 ÷ 0.7.
- k = 139.7142857...
- **Step 3:** The 139th term is therefore the last term less than 100, and the 140th term is the first term over 100.
- **Values of the Terms...**
- Term(139) = 2.2 + 0.7 × 139 = 2.2 + 97.3 = 99.5.

- Term(140) = 2.2 + 0.7 × 140 = 2.2 + 98 = 100.2.

14. Solution:

- The 4th and 1st terms are 4 - 1 = 3 positions apart.
- Common difference = (29 - 11) ÷ 3 = 18 ÷ 3 = 6.
- Zeroth term = 1st term - common difference = 11 - 6 = 5.
- Term(k) = zeroth-term + common-difference × k = 5 + 6k.
- **Step 1:** 5 + 6k = 1000.
- **Step 2:** Solve the equation...
- 6k = 1000 - 5.
- 6k = 995.
- k = 995 ÷ 6.
- k = 165.8333333...
- **Step 3:** The 165th term is therefore the last term less than 1000, and the 166th term is the first term over 1000.
- **Values of the Terms...**
- Term(165) = 5 + 6 × 165 = 5 + 990 = 995.
- Term(166) = 5 + 6 × 166 = 5 + 996 = 1001.

15. Solution:

- The 5th and 1st terms are 5 - 1 = 4 positions apart.
- Common difference = (131 - 147) ÷ 4 = -16 ÷ 4 = -4.
- Zeroth term = 1st term - common difference = 147 - (-4) = 151.
- Term(k) = zeroth-term + common-difference × k = 151 - 4k.
- **Step 1:** 151 - 4k = 0.
- **Step 2:** Solve the equation...
- -4k = 0 - 151.
- -4k = -151.
- 4k = 151.
- k = 151 ÷ 4.
- k = 37.75.
- **Step 3:** The 37th term is therefore the last term greater than 0, and the 38th term is the first term below 0.
- **Values of the Terms...**
- Term(37) = 151 - 4 × 37 = 151 - 148 = 3.
- Term(38) = 151 - 4 × 38 = 151 - 152 = -1.

16. Solution:

- Common difference = -61 - (-67) = 6.
- Zeroth term = 1st term - common difference = -67 - 6 = -73.
- Term(k) = zeroth-term + common-difference × k = -73 + 6k.
- **Step 1:** -73 + 6k = 100.
- **Step 2:** Solve the equation...
- 6k = 100 + 73.
- 6k = 173.
- k = 173 ÷ 6.
- k = 28.83333333...
- **Step 3:** The 28th term is therefore the last term less than 100, and the 29th term is the first term over 100.
- **Values of the Terms...**
- Term(28) = -73 + 6 × 28 = -73 + 168 = 95.
- Term(29) = -73 + 6 × 29 = -73 + 174 = 101.

17. Solution:

- Common difference = -11 - (-3) = -8.
- Zeroth term = 1st term - common difference = -3 - (-8) = 5.
- Term(k) = zeroth-term + common-difference × k = 5 - 8k.
- **Step 1:** 5 - 8k = -200.
- **Step 2:** Solve the equation...
- -8k = -200 - 5.
- -8k = -205.
- 8k = 205
- k = 205 ÷ 8.
- k = 25.625.
- **Step 3:** The 25th term is therefore the last term greater than -200, and the 26th term is the first term below -200.
- **Values of the Terms...**
- Term(25) = 5 - 8 × 25 = 5 - 200 = -195.
- Term(26) = 5 - 8 × 26 = 5 - 208 = -203.

18. Solution:

- Common difference = 135 - 141 = -6.
- Zeroth term = 1st term - common difference = 141 - (-6) = 147.
- Term(k) = zeroth-term + common-difference × k = 147 - 6k.
- **Step 1:** 147 - 6k = 0.

- **Step 2:** Solve the equation...
- $-6k = 0 - 147$.
- $-6k = -147$.
- $6k = 147$.
- $k = 147 \div 6$.
- $k = 24.5$.
- **Step 3:** The 24th term is therefore the last term greater than 0, and the 25th term is the first term below 0.
- **Values of the Terms...**
- $\text{Term}(24) = 147 - 6 \times 24 = 147 - 144 = 3$.
- $\text{Term}(25) = 147 - 6 \times 25 = 147 - 150 = -3$.

19. Solution:

- The 5th and 2nd terms are 5 - 2 = 3 positions apart.
- Common difference = (56 - 23) ÷ 3 = 33 ÷ 3 = 11.
- 1st term = 2nd term - common difference = 23 - 11 = 12.
- Zeroth term = 1st term - common difference = 12 - 11 = 1.
- $\text{Term}(k)$ = zeroth-term + common-difference $\times k = 1 + 11k$.
- **Step 1:** $1 + 11k = 1000$.
- **Step 2:** Solve the equation...
- $11k = 1000 - 1$.
- $11k = 999$.
- $k = 999 \div 11$.
- $k = 90.81818181...$
- **Step 3:** The 90th term is therefore the last term less than 1000, and the 91st term is the first term over 100.
- **Values of the Terms...**
- $\text{Term}(90) = 1 + 11 \times 90 = 1 + 990 = 991$.
- $\text{Term}(91) = 1 + 11 \times 91 = 1 + 1001 = 1002$.

20. Solution:

- Common difference = 104 - 112 = -8.
- Zeroth term = 1st term - common difference = 112 - (-8) = 120.
- $\text{Term}(k)$ = zeroth-term + common-difference $\times k = 120 - 8k$.
- **Step 1:** $120 - 8k = 0$.
- **Step 2:** Solve the equation...
- $-8k = 0 - 120$.
- $-8k = -120$.

- $8k = 120$.
- $k = 120 \div 8$.
- $k = 15$.
- **Step 3:** The 15th term is exactly 0. Therefore the 14th term is therefore the last term greater than 0, and the 16th term is the first term below 0.
- **Values of the Terms...**
- Term(14) = 120 - 8 × 14 = 120 - 112 = 8.
- Term(15) = 120 - 8 × 15 = 120 - 120 = 0.
- Term(16) = 120 - 8 × 16 = 120 - 128 = -8.

Chapter 3: A General Formula for the Terms in an Arithmetic Sequences

In Chapter 2, we looked at creating and using formulae to describe particular arithmetic sequences. In this chapter, we will look at a general formula which describes the terms in all arithmetic sequences.

Creating the Formula

Let us say that a is the first term of an arithmetic sequence, and d is the common difference.

In Chapter 2, we established that the zeroth term can be calculated:

- Zeroth term = 1st term - common difference.

It therefore follows:

- Zeroth term = a - d.

In Chapter 2, we also established that terms in an arithmetic sequence can be calculated:

- Term(k) = zeroth-term + common-difference × k.

Note: As mentioned in Chapter 2, in this book we use k to indicate the position (starting from 1) of a particular term in a sequence. Some other books use n for this purpose. We have avoided using n, because n is generally used (including in this book) to indicate the total number of terms in a sequence.

Substituting into this formula, we get:

- Term(k) = (a - d) + d × k.

Which can be more concisely written as:

- Term(k) = (a - d) + dk.

With some slight rearranging, this formula can also be written:

- Term(k) = $a + d(k - 1)$.

Using the Formula

Of course, this formula tends to be **not** very useful unless we know (or can find) the values of a and d. But, if we do, then can simply substitute values for a, d and k into the formula to the find the value of any term that we want.

Example: The first term in an arithmetic sequence is 23 and the common difference is 11. Using the formula, what is the 9th term of the sequence?

Solution:

- a = first term = 23.
- d = common difference = 11.
- 9th term, therefore $k = 9$.
- Using term(k) = $(a - d) + dk$: term(9) = $(23 - 11) + 11 \times 9 = 12 + 99 = 111$.

Alternate solution:

- a = first term = 23.
- d = common difference = 11.
- 9th term, therefore $k = 9$.
- Using term(k) = $a + d(k - 1)$: term(9) = $23 + 11 \times (9 - 1) = 23 + 11 \times 8 = 23 + 88 = 111$.

Example: The first term in an arithmetic sequence is 19 and the common difference is -8. Using the formula, what is the 7th term of the sequence?

Solution:

- a = first term = 19.
- d = common difference = -8.
- 7th term, therefore $k = 7$.

- Using term(k) = (a - d) + dk: term(7) = (19 - -8) + -8 × 7 = 27 - 56 = -29.

Alternate solution:

- a = first term = 19.
- d = common difference = -8.
- 7th term, therefore k = 7.
- Using term(k) = a + d(k - 1): term(7) = 19 - 8 × (7 - 1) = 19 - 8 × 6 = 19 - 48 = -29.

Example: The first term in an arithmetic sequence is -25 and the common difference is 6. Using the formula, what is the 14th term of the sequence?

Solution:

- a = first term = -25.
- d = common difference = 6.
- 14th term, therefore k = 14.
- Using term(k) = (a - d) + dk: term(14) = (-25 - 6) + 6 × 14 = -31 + 84 = 53.

Alternate solution:

- a = first term = -25.
- d = common difference = 6.
- 14th term, therefore k = 14.
- Using term(k) = a + d(k - 1): term(14) = -25 + 6 × (14 - 1) = -25 + 6 × 13 = -25 + 78 = 53.

Finding the Values of a and d

So how do we find the values of a and d of arithmetic sequence?

We could of course find the values of the first term (a) and the common difference (d) using the methods already discussed in Chapters 1 and 2, but the term formula also opens the possibility of **easily** finding their values from **any** two terms in the arithmetic sequence:

If we know the position and value of any term in an arithmetic sequence, this allows us to set-up an equation involving a and d. Knowing the position and value of a second term in the arithmetic sequence, allows us to set-up a second equation. And two equations, with two unknowns (a and d), can be solved as simultaneous equations.

Example: The 7th of an arithmetic sequence is 38. The 11th term of the 74. What are the values of the first term, and common difference? What is the formula for the k'th term of the sequence?

Solution:

- Term(7) = $a + (7 - 1)d = a + 6d = 38$.
- Term(11) = $a + (11 - 1)d = a + 10d = 74$.
- **We have simultaneous equations...**
- Equation 1: $a + 6d = 38$.
- Equation 2: $a + 10d = 74$.
- **Solve the simultaneous equations...**
- First combine the equations by doing Equation 2 - Equation 1:
- $(a + 10d) - (a + 6d) = 74 - 38$.
- $4d = 36$.
- $d = 36 \div 4$.
- $d = 9$.
- Substituting back into Equation 1...
- $a + 6 \times 9 = 38$.
- $a + 54 = 38$.
- $a = 38 - 54$.
- $a = -16$.
- **Solution...**
- First term = $a = -16$.
- Common difference = $d = 9$.
- Term(k) = $(a - d) + dk = (-16 - 9) + 9k = -25 + 9k$.

Example: The 9th of an arithmetic sequence is -3.5. The 18th term of the -10.7. What are the values of the first term, and common difference? What is the formula for the k'th term of the sequence?

Solution:

- Term(9) = $a + (9 - 1)d = a + 8d = -3.5$.
- Term(18) = $a + (18 - 1)d = a + 17d = -10.7$.

- **We have simultaneous equations...**
- Equation 1: $a + 8d = -3.5$.
- Equation 2: $a + 17d = -10.7$.
- **Solve the simultaneous equations...**
- First combine the equations by doing Equation 2 - Equation 1:
- $(a + 17d) - (a + 8d) = -10.7 - -3.5$.
- $9d = -7.2$.
- $d = -7.2 \div 9$.
- $d = -0.8$.
- Substituting back into Equation 1...
- $a + 8 \times -0.8 = -3.5$.
- $a - 6.4 = -3.5$.
- $a = -3.5 + 6.4$.
- $a = 2.9$.
- **Solution...**
- First term $= a = 2.9$.
- Common difference $= d = -0.8$.
- $\text{Term}(k) = (a - d) + dk = (2.9 - -0.8) - 0.8k = 3.7 - 0.8k$.

Example: The 1st of an arithmetic sequence is 25. The 11th term of the 95. What are the values of the first term, and common difference? What is the formula for the k'th term of the sequence?

Solution:

- $\text{Term}(1) = a + (1 - 1)d = a + 0d = a = 25$.
- $\text{Term}(11) = a + (11 - 1)d = a + 10d = 95$.
- **We have simultaneous equations...**
- Equation 1: $a = 25$.
- Equation 2: $a + 10d = 95$.
- **Solve the simultaneous equations...**
- First combine the equations by doing Equation 2 - Equation 1:
- $(a + 10d) - (a) = 95 - 25$.
- $10d = 70$.
- $d = 70 \div 10$.
- $d = 7$.
- We also already know from Equation 1...
- $a = 25$.
- **Solution...**
- First term $= a = 25$.
- Common difference $= d = 7$.

- Term(k) = ($a - d$) + dk = (25 - 7) + 7k = 18 + 7k.

Example: The 7th of an arithmetic sequence is 49. The 10th term of the 70. What are the values of the first term, and common difference? What is the formula for the k'th term of the sequence?

Solution:

- Term(7) = a + (7 - 1)d = a + 6d = 49.
- Term(10) = a + (10 - 1)d = a + 9d = 70.
- **We have simultaneous equations...**
- Equation 1: a + 6d = 49.
- Equation 2: a + 9d = 70.
- **Solve the simultaneous equations...**
- First combine the equations by doing Equation 2 - Equation 1:
- (a + 9d) - (a + 6d) = 70 - 49.
- 3d = 21.
- d = 21 ÷ 3.
- d = 7.
- Substituting back into Equation 1...
- a + 6 × 7 = 49.
- a + 42 = 49.
- a = 49 - 42.
- a = 7.
- **Solution...**
- First term = a = 7.
- Common difference = d = 7.
- Term(k) = ($a - d$) + dk = (7 - 7) + 7k = 0 + 7k = 7k.

Questions

1. The first term of an arithmetic sequence is 9 and the common difference 13. What is the 7th term of the sequence?

2. The first term of an arithmetic sequence is -5 and the common difference 7. What is the 9th term of the sequence?

3. The first term of an arithmetic sequence is 11 and the common difference -4. What is the 4th term of the sequence?

4. The first term of an arithmetic sequence is 6 and the common difference 2.5. What is the 4th term of the sequence?

5. The first term of an arithmetic sequence is -3 and the common difference -5. What is the 11th term of the sequence?

6. The first term of an arithmetic sequence is 101 and the common difference -10. What is the 10th term of the sequence?

7. The first term of an arithmetic sequence is 6 and the common difference 16. What is the 16th term of the sequence?

8. The first term of an arithmetic sequence is 3 and the common difference 7. What is the 8th term of the sequence?

9. The first term of an arithmetic sequence is 1.3 and the common difference 0.4. What is the 8th term of the sequence?

10. The first term of an arithmetic sequence is 2.1 and the common difference -0.2. What is the 7th term of the sequence?

11. The 5th of an arithmetic sequence is 14. The 10th term of the sequence is 54. What are the values of the first term, and common difference? What is the formula for the k'th term of the sequence? What is the 20th term of the sequence?

12. The 2nd of an arithmetic sequence is -4. The 51st term of the sequence is 339. What are the values of the first term, and common difference? What is the formula for the k'th term of the sequence? What is the 100th term of the sequence?

13. The 13th of an arithmetic sequence is 100. The 20th term of the sequence is 23. What are the values of the first term, and common difference? What is the formula for the k'th term of the sequence? What is the 100th term of the sequence?

14. The 7th of an arithmetic sequence is 1.9. The 13th term of the sequence is 4.3. What are the values of the first term, and common difference? What is the formula for the k'th term of the sequence? What is the 20th term of the sequence?

15. The 4th of an arithmetic sequence is -3. The 12th term of the sequence is 61. What are the values of the first term, and common difference? What is the formula for the k'th term of the sequence? What is the 20th term of the sequence?

16. The 6th of an arithmetic sequence is 7.2. The 16th term of the sequence is 15.2. What are the values of the first term, and common difference? What is the formula for the k'th term of the sequence? What is the 20th term of the sequence?

17. The 25th of an arithmetic sequence is 99. The 45th term of the sequence is 199. What are the values of the first term, and common difference? What is the formula for the k'th term of the sequence? What is the 100th term of the sequence?

18. The 1st of an arithmetic sequence is -9. The 10th term of the sequence is -90. What are the values of the first term, and common difference? What is the formula for the k'th term of the sequence? What is the 20th term of the sequence?

19. The 14th of an arithmetic sequence is 3. The 18th term of the sequence is -17. What are the values of the first term, and common difference? What is the formula for the k'th term of the sequence? What is the 20th term of the sequence?

20. The 3rd of an arithmetic sequence is -10. The 12th term of the sequence is 53. What are the values of the first term, and common difference? What is the formula for the k'th term of the sequence? What is the 20th term of the sequence?

You Can Do Math: Arithmetic Sequences

Answers to Chapter 3 Questions

1. Solution:

- a = first term = 9.
- d = common difference = 13.
- 7th term, therefore k = 7.
- Using term(k) = ($a - d$) + dk: term(7) = (9 - 13) + 13 × 7 = -4 + 91 = 87.

Alternate solution:

- a = first term = 9.
- d = common difference = 13.
- 7th term, therefore k = 7.
- Using term(k) = $a + d(k - 1)$: term(7) = 9 + 13 × (7 - 1) = 9 + 13 × 6 = 9 + 78 = 87.

2. Solution:

- a = first term = -5.
- d = common difference = 7.
- 9th term, therefore k = 9.
- Using term(k) = ($a - d$) + dk: term(9) = (-5 - 7) + 7 × 9 = -12 + 63 = 51.

Alternate solution:

- a = first term = -5.
- d = common difference = 7.
- 9th term, therefore k = 9.
- Using term(k) = $a + d(k - 1)$: term(9) = -5 + 7 × (9 - 1) = -5 + 7 × 8 = -5 + 56 = 51.

3. Solution:

- a = first term = 11.
- d = common difference = -4.
- 4th term, therefore k = 4.
- Using term(k) = ($a - d$) + dk: term(4) = (11 - -4) - 4 × 4 = 15 - 16 = -1.

Alternate solution:

- a = first term = 11.
- d = common difference = -4.
- 4th term, therefore k = 4.
- Using term(k) = $a + d(k - 1)$: term(4) = 11 - 4 × (4 - 1) = 11 - 4 × 3 = 11 - 12 = -1.

4. Solution:

- a = first term = 6.
- d = common difference = 2.5.
- 4th term, therefore k = 4.
- Using term(k) = $(a - d) + dk$: term(4) = (6 - 2.5) + 2.5 × 4 = 3.5 + 10 = 13.5.

Alternate solution:

- a = first term = 6.
- d = common difference = 2.5.
- 4th term, therefore k = 4.
- Using term(k) = $a + d(k - 1)$: term(4) = 6 + 2.5 × (4 - 1) = 6 + 2.5 × 3 = 6 + 7.5 = 13.5.

5. Solution:

- a = first term = -3.
- d = common difference = -5.
- 11th term, therefore k = 11.
- Using term(k) = $(a - d) + dk$: term(11) = (-3 - -5) - 5 × 11 = 2 - 55 = -53.

Alternate solution:

- a = first term = -3.
- d = common difference = -5.
- 11th term, therefore k = 11.
- Using term(k) = $a + d(k - 1)$: term(11) = -3 - 5 × (11 - 1) = -3 - 5 × 10 = -3 - 50 = -53.

6. Solution:

- a = first term = 101.
- d = common difference = -10.
- 10th term, therefore k = 10.

- Using term(k) = (a - d) + dk: term(10) = (101 - -10) - 10 × 10 = 111 - 100 = 11.

Alternate solution:

- a = first term = 101.
- d = common difference = -10.
- 10th term, therefore k = 10.
- Using term(k) = a + d(k - 1): term(10) = 101 - 10 × (10 - 1) = 101 - 10 × 9 = 101 - 90 = 11.

7. Solution:

- a = first term = 6.
- d = common difference = 16.
- 16th term, therefore k = 16.
- Using term(k) = (a - d) + dk: term(16) = (6 - 16) + 16 × 16 = -10 + 256 = 246.

Alternate solution:

- a = first term = 6.
- d = common difference = 16.
- 16th term, therefore k = 16.
- Using term(k) = a + d(k - 1): term(16) = 6 + 16 × (16 - 1) = 6 + 16 × 15 = 6 + 240 = 246.

8. Solution:

- a = first term = 3.
- d = common difference = 7.
- 8th term, therefore k = 8.
- Using term(k) = (a - d) + dk: term(8) = (3 - 7) + 7 × 8 = -4 + 56 = 52.

Alternate solution:

- a = first term = 3.
- d = common difference = 7.
- 8th term, therefore k = 8.
- Using term(k) = a + d(k - 1): term(8) = 3 + 7 × (8 - 1) = 3 + 7 × 7 = 3 + 49 = 52.

9. Solution:

- a = first term = 1.3.
- d = common difference = 0.4.
- 8th term, therefore k = 8.
- Using term(k) = (a - d) + dk: term(8) = (1.3 - 0.4) + 0.4 × 8 = 0.9 + 3.2 = 4.1.

Alternate solution:

- a = first term = 1.3.
- d = common difference = 0.4.
- 8th term, therefore k = 8.
- Using term(k) = a + d(k - 1): term(8) = 1.3 + 0.4 × (8 - 1) = 1.3 + 0.4 × 7 = 1.3 + 2.8 = 4.1.

10. Solution:

- a = first term = 2.1.
- d = common difference = -0.2.
- 7th term, therefore k = 7.
- Using term(k) = (a - d) + dk: term(7) = (2.1 - -0.2) - 0.2 × 7 = 2.3 - 1.4 = 0.9.

Alternate solution:

- a = first term = 2.1.
- d = common difference = -0.2.
- 7th term, therefore k = 7.
- Using term(k) = a + d(k - 1): term(7) = 2.1 - 0.2 × (7 - 1) = 2.1 - 0.2 × 6 = 2.1 - 1.2 = 0.9.

11. Solution:

- Term(5) = a + (5 - 1)d = a + 4d = 14.
- Term(10) = a + (10 - 1)d = a + 9d = 54.
- **We have simultaneous equations...**
- Equation 1: a + 4d = 14.
- Equation 2: a + 9d = 54.
- **Solve the simultaneous equations...**
- First combine the equations by doing Equation 2 - Equation 1:
- (a + 9d) - (a + 4d) = 54 - 14.
- 5d = 40.

- $d = 40 \div 5$.
- $d = 8$.
- Substituting back into Equation 1...
- $a + 4 \times 8 = 14$.
- $a + 32 = 14$.
- $a = 14 - 32$.
- $a = -18$.
- **Solution...**
- First term $= a = -18$.
- Common difference $= d = 8$.
- Term$(k) = (a - d) + dk = (-18 - 8) + 8k = -26 + 8k$.
- Term$(20) = -26 + 8 \times 20 = -26 + 160 = 134$.

12. Solution:

- Term$(2) = a + (2 - 1)d = a + d = -4$.
- Term$(51) = a + (51 - 1)d = a + 50d = 339$.
- **We have simultaneous equations...**
- Equation 1: $a + d = -4$.
- Equation 2: $a + 50d = 339$.
- **Solve the simultaneous equations...**
- First combine the equations by doing Equation 2 - Equation 1:
- $(a + 50d) - (a + d) = 339 - -4$.
- $49d = 343$.
- $d = 343 \div 49$.
- $d = 7$.
- Substituting back into Equation 1...
- $a + 7 = -4$.
- $a = -4 - 7$.
- $a = -11$.
- **Solution...**
- First term $= a = -11$.
- Common difference $= d = 7$.
- Term$(k) = (a - d) + dk = (-11 - 7) + 7k = -18 + 7k$.
- Term$(100) = -18 + 7 \times 100 = -18 + 700 = 682$.

13. Solution:

- Term$(13) = a + (13 - 1)d = a + 12d = 100$.
- Term$(20) = a + (20 - 1)d = a + 19d = 23$.

- **We have simultaneous equations...**
- Equation 1: $a + 12d = 100$.
- Equation 2: $a + 19d = 23$.
- **Solve the simultaneous equations...**
- First combine the equations by doing Equation 2 - Equation 1:
- $(a + 19d) - (a + 12d) = 23 - 100$.
- $7d = -77$.
- $d = -77 \div 7$.
- $d = -11$.
- Substituting back into Equation 1...
- $a + 12 \times -11 = 100$.
- $a - 132 = 100$.
- $a = 100 + 132$.
- $a = 232$.
- **Solution...**
- First term $= a = 232$.
- Common difference $= d = -11$.
- $\text{Term}(k) = (a - d) + dk = (232 - -11) - 11k = 243 - 11k$.
- $\text{Term}(100) = 243 - 11 \times 100 = 243 - 1100 = -857$.

14. Solution:

- $\text{Term}(7) = a + (7 - 1)d = a + 6d = 1.9$.
- $\text{Term}(13) = a + (13 - 1)d = a + 12d = 4.3$.
- **We have simultaneous equations...**
- Equation 1: $a + 6d = 1.9$.
- Equation 2: $a + 12d = 4.3$.
- **Solve the simultaneous equations...**
- First combine the equations by doing Equation 2 - Equation 1:
- $(a + 12d) - (a + 6d) = 4.3 - 1.9$.
- $6d = 2.4$.
- $d = 2.4 \div 6$.
- $d = 0.4$.
- Substituting back into Equation 1...
- $a + 0.4 \times 6 = 1.9$.
- $a + 2.4 = 1.9$.
- $a = 1.9 - 2.4$.
- $a = -0.5$.
- **Solution...**
- First term $= a = -0.5$.
- Common difference $= d = 0.4$.

- Term(k) = ($a - d$) + dk = (-0.5 - 0.4) + 0.4k = -0.9 + 0.4k.
- Term(20) = -0.9 + 20 × 0.4 = -0.9 + 8 = 7.1.

15. Solution:

- Term(4) = a + (4 - 1)d = a + 3d = -3.
- Term(12) = a + (12 - 1)d = a + 11d = 61.
- **We have simultaneous equations...**
- Equation 1: a + 3d = -3.
- Equation 2: a + 11d = 61.
- **Solve the simultaneous equations...**
- First combine the equations by doing Equation 2 - Equation 1:
- (a + 11d) - (a + 3d) = 61 - -3.
- 8d = 64.
- d = 64 ÷ 8.
- d = 8.
- Substituting back into Equation 1...
- a + 8 × 3 = -3.
- a + 24 = -3.
- a = -3 - 24.
- a = -27.
- **Solution...**
- First term = a = -27.
- Common difference = d = 8.
- Term(k) = ($a - d$) + dk = (-27 - 8) + 8k = -35 + 8k.
- Term(20) = -35 + 8 × 20 = -35 + 160 = 125.

16. Solution:

- Term(6) = a + (6 - 1)d = a + 5d = 7.2.
- Term(16) = a + (16 - 1)d = a + 15d = 15.2.
- **We have simultaneous equations...**
- Equation 1: a + 5d = 7.2.
- Equation 2: a + 15d = 15.2.
- **Solve the simultaneous equations...**
- First combine the equations by doing Equation 2 - Equation 1:
- (a + 15d) - (a + 5d) = 15.2 - 7.2.
- 10d = 8.
- d = 8 ÷ 10.
- d = 0.8.
- Substituting back into Equation 1...
- a + 0.8 × 5 = 7.2.

- $a + 4 = 7.2$.
- $a = 7.2 - 4$.
- $a = 3.2$.
- **Solution...**
- First term $= a = 3.2$.
- Common difference $= d = 0.8$.
- Term$(k) = (a - d) + dk = (3.2 - 0.8) + 0.8k = 2.4 + 0.8k$.
- Term$(20) = 2.4 + 0.8 \times 20 = 2.4 + 16 = 18.4$.

17. Solution:

- Term$(25) = a + (25 - 1)d = a + 24d = 99$.
- Term$(45) = a + (45 - 1)d = a + 44d = 199$.
- **We have simultaneous equations...**
- Equation 1: $a + 24d = 99$.
- Equation 2: $a + 44d = 199$.
- **Solve the simultaneous equations...**
- First combine the equations by doing Equation 2 - Equation 1:
- $(a + 44d) - (a + 24d) = 199 - 99$.
- $20d = 100$.
- $d = 100 \div 20$.
- $d = 5$.
- Substituting back into Equation 1...
- $a + 5 \times 24 = 99$.
- $a + 120 = 99$.
- $a = 99 - 120$.
- $a = -21$.
- **Solution...**
- First term $= a = -21$.
- Common difference $= d = 5$.
- Term$(k) = (a - d) + dk = (-21 - 5) + 5k = -26 + 5k$.
- Term$(100) = -26 + 5 \times 100 = -26 + 500 = 474$.

18. Solution:

- Term$(1) = a + (1 - 1)d = a + 0d = a = -9$.
- Term$(10) = a + (10 - 1)d = a + 9d = -90$.
- **We have simultaneous equations...**
- Equation 1: $a = -9$.
- Equation 2: $a + 9d = 90$.

- **Solve the simultaneous equations...**
- First combine the equations by doing Equation 2 - Equation 1:
- $(a + 9d) - (a) = -90 - -9$.
- $9d = -81$.
- $d = -81 \div 9$.
- $d = -9$.
- We also already know from Equation 1...
- $a = -9$.
- **Solution...**
- First term = $a = -9$.
- Common difference = $d = -9$.
- Term(k) = $(a - d) + dk = (-9 - -9) - 9k = 0 - 9k = -9k$.
- Term(20) = $-9 \times 20 = -180$.

19. Solution:

- Term(14) = $a + (14 - 1)d = a + 13d = 3$.
- Term(18) = $a + (18 - 1)d = a + 17d = -17$.
- **We have simultaneous equations...**
- Equation 1: $a + 13d = 3$.
- Equation 2: $a + 17d = -17$.
- **Solve the simultaneous equations...**
- First combine the equations by doing Equation 2 - Equation 1:
- $(a + 17d) - (a + 13d) = -17 - 3$.
- $4d = -20$.
- $d = -20 \div 4$.
- $d = -5$.
- Substituting back into Equation 1...
- $a + -5 \times 13 = 3$.
- $a - 65 = 3$.
- $a = 3 + 65$.
- $a = 68$.
- **Solution...**
- First term = $a = 68$.
- Common difference = $d = -5$.
- Term(k) = $(a - d) + dk = (68 - -5) - 5k = 73 - 5k$.
- Term(20) = $73 - 5 \times 20 = 73 - 100 = -27$.

20. Solution:

- Term(3) = $a + (3 - 1)d = a + 2d = -10$.
- Term(12) = $a + (12 - 1)d = a + 11d = 53$.
- **We have simultaneous equations…**
- Equation 1: $a + 2d = -10$.
- Equation 2: $a + 11d = 53$.
- **Solve the simultaneous equations…**
- First combine the equations by doing Equation 2 - Equation 1:
- $(a + 11d) - (a + 2d) = 53 - -10$.
- $9d = 63$.
- $d = 63 \div 9$.
- $d = 7$.
- Substituting back into Equation 1…
- $a + 7 \times 2 = -10$.
- $a + 14 = -10$.
- $a = -10 - 14$.
- $a = -24$.
- **Solution…**
- First term = $a = -24$.
- Common difference = $d = 7$.
- Term(k) = $(a - d) + dk = (-24 - 7) + 7k = -38 + 7k$.
- Term(20) = $-31 + 7 \times 20 = -31 + 140 = 109$.

Chapter 4: Calculating the Sum and Mean of an Arithmetic Sequence

In Chapter 3, we looked at a general formula which can be used to calculate the value of any term of any arithmetic sequence from the values of the first term (a), the common difference (d), and the position of the desired term (k). We will now formulae for calculating the sum and mean of any arithmetic sequence based on the values of the first term (a), the common difference (d), and the total number of terms in the arithmetic sequence (n).

Gauss at Elementary School

Johann Carl Friedrich Gauss (April 30th, 1777 to February 23rd, 1855) was one of the most important mathematicians in history. He is sometimes referred to as *princeps mathematicorum* (which is Latin for "the foremost of mathematicians") or as "the greatest mathematician since antiquity".

Johann Carl Friedrich Gauss:

There are many and amusing stories about Gauss, including that at age 3, he spotted an error that his father had made on paper while preparing his finances, and immediately mentally calculated the correction. It is also claimed that when interrupted while working on a math problem to be told that his wife was dying, Gauss responded "Tell her to wait a moment till I'm done." Another Gauss story is about summing an arithmetic sequence...

According to legend, when in elementary school (primary school), Gauss misbehaved and was given the task of adding the numbers from 1 to 100 as a punishment. Gauss however produced the correct answer within seconds to the amazement of his teacher.

He did this by noticing that you could pair up the terms; the 1st term could be paired up with the last (100th) term, the 2nd term with the second to last term (99th), and every pair of terms would always add to the same value (101). Since there are 100 terms in total, there would be 100 ÷ 2 = 50 pairs, and hence the sum of all the terms is 50 × 101 = 5050.

Generating a Formula for the Sum of Arithmetic Sequence

Here's essentially the same argument presented in a slightly different way:

1st term	2nd term	...etc...	2nd to last term	Last (100th) term
1	2		99	100
Last (100th) term	**2nd to last term**	**...etc...**	**2nd term**	**1st term**
100	99		2	1
Total of Pair	**Total of Pair**	**...etc...**	**Total of Pair**	**Total of Pair**
1 + 100 = 101	2 + 99 = 101		99 + 2 = 101	100 + 1 = 101

- In the first column of the table, the 1st term of the sequence is paired with the last (100th) term of the sequence. The total that we get from adding this pair of terms is 101.
- In the second column of the table, the 2nd term of the sequence is paired with the 2nd to last term of the sequence. The total that we get from adding this pair of terms is again 101.
- We then keep going through all the terms in the sequence, always pairing and always finding the total for every pair is 100.
- At the end of the process, we have 100 pairs, but we have used each term twice. The total for 100 pairs (twice the sum of the sequence) is therefore 100 × 101 = 10100.
- The sum of the sequence is therefore 10100 ÷ 2 = 5050.

We can generalize this process using the term formula that we used in Chapter 3, to create a formula for the sum of any arithmetic sequence.

You will recall this formula from Chapter 3:

- term(k) = $a + d(k - 1)$.

So, if we say there are a total of n terms in an arithmetic sequence, we get:

- term(1) = a.
- term(2) = $a + d$.
- term(2) = $a + 2d$.
- ...etc...
- Third to last term = term($n - 2$) = $a + d(n - 3)$
- Second to last term = term($n - 1$) = $a + d(n - 2)$
- Last term = term(n) = $a + d(n - 1)$

So, if we use the above formulae with the pairing-up method that we described earlier, we get:

1st term	2nd term	...etc...	2nd to last term	Last (n'th) term
a	$a + d$		$a + d(n - 2)$	$a + d(n - 1)$
Last (n'th) term	2nd to last term	...etc...	2nd term	1st term
$a + d(n - 1)$	$a + d(n - 2)$		$a + d$	a
Total of Pair	Total of Pair	...etc...	Total of Pair	Total of Pair
$2a + d(n - 1)$	$2a + d(n - 1)$		$2a + d(n - 1)$	$2a + d(n - 1)$

- In the first column of the table, the 1st term of the sequence is paired with the last (n'th) term of the sequence. The total that we get from adding this pair of terms is $2a + d(n - 1)$.
- In the second column of the table, the 2nd term of the sequence is paired with the 2nd to last term of the sequence. The total that we get from adding this pair of terms is again $2a + d(n - 1)$.
- We then keep going through all the terms in the sequence, always pairing and always finding the total for every pair is $2a + d(n - 1)$.
- At the end of the process, we have n pairs, but we have used each term twice. The total for n pairs (twice the sum of the sequence) is therefore $n \times (2a + d(n - 1))$.
- The sum of the sequence is therefore $(n \div 2) \times (2a + d(n - 1))$. This can be simply written as $S = (n \div 2)(2a + d(n - 1))$.
- Finally, we should mention that some people like to put a subscript after the S to indicate how many terms are being summed, so S_n would mean the sum of the first n terms of an arithmetic sequence, and S_{49} would mean the sum of the 49 terms, and so on.

To summarize:

- Sum of arithmetic sequence = $S_n = (n \div 2)(2a + d(n - 1))$.

Note: Because of the amusing anecdote about his experiences in elementary school, many people wrongly attribute the discovery of this formula to Gauss. While it is true that Gauss seems to have independently discovered it (or a variant of it) at a young age, the formula was actually known centuries earlier. The earliest-known record of this formula is in the writings of the classical Indian astronomer-mathematician Aryabhata (476 AD to 550 AD).

Statue of Aryabhata in Pune, India:

Using the Sum Formula

You can use the sum formula to calculate the sum of any arithmetic sequence – provided you know the first term, common difference and number of terms. Conversely, if you know the sum of the sequence and any two out of the first term, common difference, and number of terms, you can find the third. One thing to be aware of though, is if calculating the number of terms in the sequence, you will need to solve a **quadratic equation** and reject any negative answer (since the number of terms in the sequence can **not** be negative).

Let's try some examples...

Example: An arithmetic sequence contains 20 terms. The first term is 7. The common difference is 11. What is the sum of the sequence?

Solution:

- n = Number of terms = 20.
- a = First term = 7.
- d = Common difference = 11.
- S_{20} = Sum = $(n \div 2)(2a + d(n - 1))$ = $(20 \div 2) \times (2 \times 7 + 11 \times (20 - 1))$ = $10 \times (14 + 11 \times 19)$ = $10 \times (14 + 209)$ = 10×223 = 2230.

Example: An arithmetic sequence contains 24 terms. If the first term in the sequence is 3, and the sum of all the terms in the sequence is 154.8, what is the common difference?

Solution:

- n = Number of terms = 24.
- a = First term = 3.
- S_{24} = Sum = $(n \div 2)(2a + d(n - 1))$ = $(24 \div 2) \times (2 \times 3 + d \times (24 - 1))$ = 154.8.
- **So, we need to solve this equation:**
- $(24 \div 2) \times (2 \times 3 + d \times (24 - 1))$ = 154.8.
- **Solving...**
- $12 \times (6 + 23d)$ = 154.8.
- $6 + 23d$ = $154.8 \div 12$.
- $6 + 23d$ = 12.9.
- $23d$ = 12.9 - 6.
- $23d$ = 6.9.
- d = $6.9 \div 23$.
- d = 0.3.
- Common difference = d = 0.3.

Example: The sum of the first 32 terms of an arithmetic sequence is -880. Given that the common difference is -5, what is the first term in the sequence?

Solution:

- n = Number of terms = 32.
- d = Common difference = -5.
- S_{32} = Sum = $(n \div 2)(2a + d(n - 1)) = (32 \div 2) \times (2a - 5 \times (32 - 1)) = -880$.
- **So, we need to solve this equation:**
- $(32 \div 2) \times (2a - 5 \times (32 - 1)) = -880$.
- **Solving...**
- $16 \times (2a - 5 \times (32 - 1)) = -880$.
- $(2a - 5 \times (32 - 1)) = -880 \div 16$.
- $(2a - 5 \times (32 - 1)) = -55$.
- $(2a - 5 \times 31) = -55$.
- $2a - 155 = -55$.
- $2a = -55 + 155$.
- $2a = 100$.
- $a = 100 \div 2$.
- $a = 50$.
- First term = $a = 50$.

Example: The first term of an arithmetic sequence is 5. The common difference is 3. The sum of the sequence is 185. How many terms are in the sequence?

Solution:

- a = First term = 5.
- d = Common difference = 3.
- S_n = Sum = $185 = (n \div 2)(2a + d(n - 1))$.
- **Solving the equation...**
- $185 = (n \div 2) \times (2 \times 5 + 3 \times (n - 1))$.
- $185 = (n \div 2) \times (10 + 3n - 3)$.
- $185 = (n \div 2) \times (3n + 7)$.
- $370 = n \times (3n + 7)$.
- $370 = 3n^2 + 7n$.

- $0 = 3n^2 + 7n - 370$. ← Note: **Quadratic equation**! Solve using the quadratic formula or by factorizing.
- $0 = (3n + 37)(n - 10)$.
- $n = -37/3$ or $n = 10$.
- **Solution...**
- Since n can **not** be negative...
- Number of terms = $n = 10$.

Finding the Sum of Ranges of Terms within an Arithmetic Sequence

Sometimes you may wish to find the sum of a range of terms in the middle of an arithmetic sequence. This can be most easily achieved by calculating a sum from the 1st term to the end of the desired range of terms, separately calculating the sum of the terms before the desired range of terms, and then subtracting the latter from the former.

For example, suppose you wished to find the sum of terms from the 7th to the 23rd position inclusive. You would first find the sum from the 1st to the 23rd term inclusive, separately find the sum from the 1st to the 6th term inclusive, and then subtract the latter from the former.

You can build on this process if you ever need to find the sum of multiple ranges of terms; simply calculate the sum of each range separately, and then add these sums together.

Example: The first term in arithmetic sequence is 10, the common difference is 3. What is the sum of the 7th to 23rd terms inclusive?

Solution:

- a = First term = 10.
- d = Common difference = 3.
- **Calculate the sum of terms 1 to 23, using $n = 23$:**
- Sum(1 to 23) = S_{23} = $(n \div 2)(2a + d(n - 1))$ = $(23 \div 2) \times (2 \times 10 + 3 \times (23 - 1))$ = $11.5 \times (20 + 3 \times 22)$ = $11.5 \times 86 = 989$.
- **Calculate the sum of terms 1 to 6, using $n = 6$:**
- Sum(1 to 6) = S_6 = $(n \div 2)(2a + d(n - 1))$ = $(6 \div 2) \times (2 \times 10 + 3 \times (6 - 1))$ = $3 \times (20 + 3 \times 5)$ = $3 \times 35 = 105$.

- **Calculate the sum of terms 7 to 23, by subtracting the sums:**
- Sum(7 to 23) = Sum(1 to 23) - Sum(1 to 6) = 989 - 105 = 884.

Example: The first term in arithmetic sequence is 5, the common difference is 4. What is the sum of the 9th to 19th terms (inclusive) together with the 29th to 39th terms (inclusive)?

Solution:

- a = First term = 5.
- d = Common difference = 4.
- **Calculate the sum of terms 1 to 19, using $n = 19$:**
- Sum(1 to 19) = S_{19} = ($n \div 2$)($2a + d(n - 1)$) = ($19 \div 2$) × ($2 \times 5 + 4 \times (19 - 1)$) = 9.5 × (10 + 4 × 18) = 9.5 × 82 = 779.
- **Calculate the sum of terms 1 to 8, using $n = 8$:**
- Sum(1 to 8) = S_8 = ($n \div 2$)($2a + d(n - 1)$) = ($8 \div 2$) × ($2 \times 5 + 4 \times (8 - 1)$) = 4 × (10 + 4 × 7) = 4 × 38 = 152.
- **Calculate the sum of terms 9 to 19, by subtracting the sums:**
- Sum(9 to 19) = Sum(1 to 19) - Sum(1 to 8) = 779 - 152 = 627.
- **Calculate the sum of terms 1 to 39, using $n = 39$:**
- Sum(1 to 39) = S_{39} = ($n \div 2$)($2a + d(n - 1)$) = ($39 \div 2$) × ($2 \times 5 + 4 \times (39 - 1)$) = 19.5 × (10 + 4 × 38) = 19.5 × 162 = 3159.
- **Calculate the sum of terms 1 to 28, using $n = 28$:**
- Sum(1 to 28) = S_{28} = ($n \div 2$)($2a + d(n - 1)$) = ($28 \div 2$) × ($2 \times 5 + 4 \times (28 - 1)$) = 14 × (10 + 4 × 27) = 4 × 38 = 1652.
- **Calculate the sum of terms 29 to 39, by subtracting the sums:**
- Sum(29 to 39) = Sum(1 to 39) - Sum(1 to 28) = 3159 - 1652 = 1507.
- **Add the sum of terms 9 to 19 with the sum of terms 29 to 39:**
- Answer = Sum(9 to 19) + Sum(29 to 39) = 627 + 1507 = 2134.

Generating a Formula for the Arithmetic Mean of an Arithmetic Sequence

The arithmetic mean of any list of numbers is simply the total of all the numbers divided by how many numbers there are. In the case of an arithmetic sequence, the arithmetic mean can therefore be calculated:

- Arithmetic mean = $S_n \div n$.

You can use this formula to calculate the arithmetic mean of an arithmetic sequence – provided you know the sum and number of terms. Conversely, if you know the arithmetic mean and the number of terms then you can find the sum, of if you know the arithmetic mean and the sum then you can find the number of terms.

If we substitute in the formula for Sn, this becomes:

- Arithmetic mean = $(n \div 2)(2a + d(n - 1)) \div n$.

Which simplifies to:

- Arithmetic mean = $\frac{1}{2}(2a + d(n - 1))$.

You can use this formula to calculate the arithmetic mean of an arithmetic sequence – provided you know the first term, common difference and number of terms. Conversely, if you know the arithmetic mean of the sequence and any two out of the first term, common difference, and number of terms, you can find the third.

Example: An arithmetic sequence of 24 terms begins with a first term of 9 and has a common difference of 0.5. What is the arithmetic mean of the terms?

Solution:

- n = Number of terms = 24.
- a = First term = 9.
- d = Common difference = 0.5.
- Arithmetic mean = $\frac{1}{2}(2a + d(n - 1)) = \frac{1}{2} \times (2 \times 9 + 0.5 \times (24 - 1)) = \frac{1}{2} \times (18 + 0.5 \times 23) = \frac{1}{2} \times (18 + 11.5) = \frac{1}{2} \times 29.5 = 14.75$.

Example: The arithmetic mean of an arithmetic sequence is 40. If the first term is 13 and the common difference is 3, how many terms are in the sequence?

Solution:

- a = First term = 13.

- d = Common difference = 3.
- Arithmetic mean = ½($2a + d(n - 1)$) = ½ × (2 × 13 + 3 × ($n - 1$)) = 40.
- **Solving the equation...**
- ½ × (2 × 13 + 3 × ($n - 1$)) = 40.
- (2 × 13 + 3 × ($n - 1$)) = 80
- (26 + 3n - 3) = 80.
- 3n + 23 = 80.
- 3n = 80 - 23.
- 3n = 57.
- n = 57 ÷ 3.
- n = 19.
- **Solution...**
- Number of terms = n = 19.

Simultaneous Equations Involving the Sum and/or Mean

In Chapter 3, we saw that it was possible to find both the first term and the common difference of any arithmetic sequence if you know at least two terms by using the simultaneous equations. Similarly, it is possible to use simultaneous equations to find out more about a sequence, if you know the sum/mean of the sequence, or the sums/means of different ranges of terms within the sequence.

Example: An athlete is training for 20 days. Each day he does more push-ups, increasing the number compared to the previous day by a constant amount. On the last day of training, the athlete does 120 push-ups and calculates that he has done a total of 1450 push-ups over the entire training-period. How many push-ups did he do on the first day? What was the daily increase in the number of push-ups?

Solution:

- n = Number of terms = 20.
- S_{20} = Sum = (n ÷ 2)($2a + d(n - 1)$) = (20 ÷ 2) × ($2a + d(20 - 1)$) = 10 × ($2a + 19d$) = 1450.
- Term(20) = a + (20 - 1)d = a + 19d = 120.
- **We have simultaneous equations...**
- Equation 1: 10 × ($2a + 19d$) = 1450.
- Equation 2: a + 19d = 120.
- **Now solve the simultaneous equations...**
- Rearrange Equation 1:
- 10 × ($2a + 19d$) = 1450.
- $2a + 19d$ = 1450 ÷ 10.

- $2a + 19d = 145$.
- $a + a + 19d = 145$.
- Since we know from Equation 2 that $a + 19d = 120$, we can substitute 120 for $a + 19d$ in the above:
- $a + 120 = 145$.
- $a = 145 - 120$.
- $a = 25$.
- Substitute the value of a back into Equation 2:
- $25 + 19d = 120$.
- $19d = 120 - 25$.
- $19d = 95$.
- $d = 95 \div 19$.
- $d = 5$.
- **Answer:**
- First day push-ups = First term = $a = 25$.
- Daily increase in push-ups = Common difference = $d = 5$.

Example: The sum of the first 10 terms in an arithmetic sequence is 50. The sum of the first 20 terms in the same sequence is -100. What are the first term and the common difference of the sequence?

Solution:

- $S_{10} = (n \div 2)(2a + d(n - 1)) = (10 \div 2) \times (2a + d \times (10 - 1)) = 5(2a + 9d) = 50$.
- $S_{20} = (n \div 2)(2a + d(n - 1)) = (20 \div 2) \times (2a + d \times (20 - 1)) = 10(2a + 19d) = -100$.
- **We have simultaneous equations...**
- Equation 1: $5 \times (2a + 9d) = 50$.
- Equation 2: $10 \times (2a + 19d) = -100$.
- We can simplify Equation 1:
- $2a + 9d = 50 \div 5$.
- $2a + 9d = 10$. ← Let's call this Equation 3.
- We can simplify Equation 2:
- $2a + 19d = -100 \div 10$.
- $2a + 19d = -10$. ← Let's call this Equation 4.
- **Solve the simultaneous equations...**
- First combine the equations by doing Equation 4 - Equation 1:
- $(2a + 19d) - (2a + 9d) = -10 - 10$.
- $10d = -20$.
- $d = -20 \div 10$.
- $d = -2$.

- Substituting back into Equation 3...
- $2a + 9 \times -2 = 10$.
- $2a - 18 = 10$.
- $2a = 10 + 18$.
- $2a = 28$.
- $a = 28 \div 2$.
- $a = 14$.
- **Solution...**
- First term = $a = 14$.
- Common difference = $d = -2$.

Example: The last term in an arithmetic sequence is 70, the arithmetic mean of the sequence is 37, and the sum of all the terms in the sequence is 444. What is the first term of the sequence? What is the common difference? How many terms are in the sequence?

Solution:

- We begin with the following simultaneous equations:
- Equation 1 (last term): $a + d(n - 1) = 70$.
- Equation 2 (arithmetic mean): $\frac{1}{2}(2a + d(n - 1)) = 37$.
- Equation 3 (sum of the series): $(n \div 2)(2a + d(n - 1)) = 444$.
- **Solve the simultaneous equations...**
- We can substitute Equation 2 into Equation 1, and then solve to find a:
- $\frac{1}{2}(a + 70) = 37$.
- $a + 70 = 37 \times 2$.
- $a + 70 = 74$.
- $a = 74 - 70$.
- $a = 4$.
- If we do Equation 3 ÷ Equation 2, we get:
- $n = 444 \div 37$. ← If you can't see why we get n on the left-hand side, it might help to remember the arithmetic mean is the sum divided by n. In other words, $37 = 444 \div n$.
- $n = 12$.
- We can now substitute the values of a and n into Equation 1 and solve to find d:
- $4 + d(12 - 1) = 70$.
- $4 + 11d = 70$.
- $11d = 70 - 4$.
- $11d = 66$.
- $d = 66 \div 11$.
- $d = 6$.
- **Solution...**

- First term = a = 4.
- Common difference = d = 6.
- Number of terms = n = 12.

Questions

1. What is the sum of all the integers (whole number) from 1 to 1000 inclusive?

2. The first term of an arithmetic sequence is 11 and the common difference is -4. What is the sum of the first 40 terms of the sequence?

3. The first term of an arithmetic sequence is 19 and the common difference is 3. What is the sum of the first 25 terms of the sequence?

4. The first term of an arithmetic sequence is 2.1 and common difference is 0.4. What is the sum of the first 18 terms of the sequence?

5. The first term of an arithmetic sequence is -12 and the common difference is 5. What is the sum of the first 42 terms of the sequence?

6. The first term of an arithmetic sequence is 13 and the common difference is 4. What is the sum of terms 11 to 20 (inclusive) of the sequence?

7. The first term of an arithmetic sequence is 56 and the common difference is -7. What is the sum of terms 13 to 24 (inclusive) of the sequence?

8. The first term of an arithmetic sequence is -25 and the common difference is 3. What is the sum of terms 5 to 10 (inclusive) of the sequence?

9. The first term of an arithmetic sequence is 8 and the common difference is 11. What is the sum of terms 7 to 14 (inclusive) of the sequence?

10. The first term of an arithmetic sequence is 5 and the common difference is 1.5. What is the sum of terms 6 to 10 (inclusive) of the sequence?

11. An arithmetic sequence has a first term of 7 and a common difference of 4. How many terms would need to be added to find reach a total of 2772.

12. An arithmetic sequence has a first term of 10 and a common difference of 6. How many terms would need to be added to find reach a total of 1896.

13. An arithmetic sequence has a first term of 5 and a common difference of 7. How many terms would need to be added to find reach a total of 1727.

14. An arithmetic sequence has a first term of -8 and a common difference of 3. How many terms would need to be added to find reach a total of 161.

15. An arithmetic sequence has a first term of 10 and a common difference of 11. How many terms would need to be added to find reach a total of 388.

16. An arithmetic sequence contains 14 terms. The last term is 55 and the sum of the terms is 406. What is the first term and the common difference?

17. The sum of the first 12 terms of an arithmetic sequence is 246. The sum of the first 24 terms of the same sequence is 924. What is the first term and the common difference?

18. An arithmetic sequence has a common difference of 7. The last term in the sequence 148. The arithmetic mean of the terms in the sequence is 81.5. What is the first term in the sequence? How many terms are in the sequence?

19. An arithmetic sequence contains 24 terms. The last term in the sequence is -99. The arithmetic mean of the terms is -41.5.   What is the first term and the common difference?

20. An arithmetic sequence contains 100 terms. The last term in the sequence is 41.3. The sum of the terms is 2150.   What is the first term and the common difference?

Answers to Chapter 4 Questions

1. Solution:

- n = Number of terms = 1000.
- a = First term = 1.
- d = Common difference = 1.
- S_{1000} = Sum = $(n \div 2)(2a + d(n - 1)) = (1000 \div 2) \times (2 \times 1 + 1 \times (1000 - 1)) = 500 \times (2 + 1 \times 999) = 500 \times (2 + 999) = 500 \times 1001 = 500500$.

2. Solution:

- n = Number of terms = 40.
- a = First term = 11.
- d = Common difference = -4.
- S_{40} = Sum = $(n \div 2)(2a + d(n - 1)) = (40 \div 2) \times (2 \times 11 - 4 \times (40 - 1)) = 20 \times (22 - 4 \times 39) = 20 \times (22 - 156) = 20 \times -134 = -2680$.

3. Solution:

- n = Number of terms = 25.
- a = First term = 19.
- d = Common difference = 3.
- S_{25} = Sum = $(n \div 2)(2a + d(n - 1)) = (25 \div 2) \times (2 \times 19 + 3 \times (25 - 1)) = 12.5 \times (38 + 3 \times 24) = 12.5 \times (38 + 72) = 12.5 \times 110 = 1375$.

4. Solution:

- n = Number of terms = 18.
- a = First term = 2.1.
- d = Common difference = 0.4.
- S_{18} = Sum = $(n \div 2)(2a + d(n - 1)) = (18 \div 2) \times (2 \times 2.1 + 0.4 \times (18 - 1)) = 9 \times (4.2 + 0.4 \times 17) = 9 \times (4.2 + 6.8) = 9 \times 11 = 99$.

5. Solution:

- n = Number of terms = 42.
- a = First term = -12.
- d = Common difference = 5.

- S_{42} = Sum = $(n \div 2)(2a + d(n - 1))$ = $(42 \div 2) \times (2 \times -12 + 5 \times (42 - 1))$ = $21 \times (-24 + 205)$ = $21 \times (-24 + 205)$ = 21×181 = 3801.

6. Solution:

- a = First term = 13.
- d = Common difference = 4.
- **Calculate the sum of terms 1 to 20, using n = 20:**
- Sum(1 to 20) = S_{20} = $(n \div 2)(2a + d(n - 1))$ = $(20 \div 2) \times (2 \times 13 + 4 \times (20 - 1))$ = $10 \times (26 + 4 \times 19)$ = 10×102 = 1020.
- **Calculate the sum of terms 1 to 10, using n = 10:**
- Sum(1 to 10) = S_{10} = $(n \div 2)(2a + d(n - 1))$ = $(10 \div 2) \times (2 \times 13 + 4 \times (10 - 1))$ = $5 \times (26 + 4 \times 9)$ = 5×62 = 310.
- **Calculate the sum of terms 11 to 20, by subtracting the sums:**
- Sum(11 to 20) = Sum(1 to 20) - Sum(1 to 10) = 1020 - 310 = 710.

7. Solution:

- a = First term = 56.
- d = Common difference = -7.
- **Calculate the sum of terms 1 to 24, using n = 24:**
- Sum(1 to 24) = S_{24} = $(n \div 2)(2a + d(n - 1))$ = $(24 \div 2) \times (2 \times 56 - 7 \times (24 - 1))$ = $12 \times (112 - 7 \times 23)$ = 12×-49 = -588.
- **Calculate the sum of terms 1 to 12, using n = 12:**
- Sum(1 to 12) = S_{12} = $(n \div 2)(2a + d(n - 1))$ = $(12 \div 2) \times (2 \times 56 - 7 \times (12 - 1))$ = $6 \times (112 - 7 \times 11)$ = 6×35 = 210.
- **Calculate the sum of terms 7 to 23, by subtracting the sums:**
- Sum(13 to 24) = Sum(1 to 24) - Sum(1 to 12) = -588 - 210 = -798.

8. Solution:

- a = First term = -25.
- d = Common difference = 3.
- **Calculate the sum of terms 1 to 10, using n = 10:**
- Sum(1 to 10) = S_{10} = $(n \div 2)(2a + d(n - 1))$ = $(10 \div 2) \times (2 \times -25 + 3 \times (10 - 1))$ = $5 \times (-50 + 3 \times 9)$ = 5×-23 = -115.
- **Calculate the sum of terms 1 to 4, using n = 4:**
- Sum(1 to 4) = S_4 = $(n \div 2)(2a + d(n - 1))$ = $(4 \div 2) \times (2 \times -25 + 3 \times (4 - 1))$ = $2 \times (-50 + 3 \times 3)$ = 2×-41 = -82.
- **Calculate the sum of terms 5 to 10, by subtracting the sums:**

- Sum(5 to 10) = Sum(1 to 10) - Sum(1 to 4) = -115 - -82 = -33.

9. Solution:

- a = First term = 8.
- d = Common difference = 11.
- **Calculate the sum of terms 1 to 14, using $n = 14$:**
- Sum(1 to 14) = S_{14} = ($n \div 2$)($2a + d(n - 1)$) = ($14 \div 2$) × ($2 \times 8 + 11 \times (14 - 1)$) = $7 \times (16 + 11 \times 13)$ = $7 \times 159 = 1113$.
- **Calculate the sum of terms 1 to 6, using $n = 6$:**
- Sum(1 to 6) = S_6 = ($n \div 2$)($2a + d(n - 1)$) = ($6 \div 2$) × ($2 \times 8 + 11 \times (6 - 1)$) = $3 \times (16 + 11 \times 5)$ = $3 \times 71 = 213$.
- **Calculate the sum of terms 7 to 14, by subtracting the sums:**
- Sum(7 to 14) = Sum(1 to 14) - Sum(1 to 6) = 1113 - 213 = 900.

10. Solution:

- a = First term = 5.
- d = Common difference = 1.5.
- **Calculate the sum of terms 1 to 10, using $n = 10$:**
- Sum(1 to 10) = S_{10} = ($n \div 2$)($2a + d(n - 1)$) = ($10 \div 2$) × ($2 \times 5 + 1.5 \times (10 - 1)$) = $5 \times (10 + 1.5 \times 9)$ = $5 \times 23.5 = 117.5$.
- **Calculate the sum of terms 1 to 5, using $n = 5$:**
- Sum(1 to 5) = S_5 = ($n \div 2$)($2a + d(n - 1)$) = ($5 \div 2$) × ($2 \times 5 + 1.5 \times (5 - 1)$) = $2.5 \times (10 + 1.5 \times 4)$ = $2.5 \times 16 = 40$.
- **Calculate the sum of terms 7 to 23, by subtracting the sums:**
- Sum(6 to 10) = Sum(1 to 10) - Sum(1 to 5) = 117.5 - 40 = 77.5.

11. Solution:

- a = First term = 7.
- d = Common difference = 4.
- S_n = Sum = 2772 = ($n \div 2$)($2a + d(n - 1)$).
- **Solving the equation...**
- 2772 = ($n \div 2$) × ($2 \times 7 + 4 \times (n - 1)$).
- 2772 = ($n \div 2$) × ($14 + 4n - 4$).
- 2772 = ($n \div 2$) × ($4n + 10$).
- 5544 = $n \times (4n + 10)$.
- 5544 = $4n^2 + 10n$.

- $0 = 4n^2 + 10n - 5544$. ← Note: **Quadratic equation**! Solve using the quadratic formula or by factorizing.
- First simplify by dividing both sides by 2...
- $0 = 2n^2 + 5n - 2772$.
- Then factorize...
- $0 = (2n + 77)(n - 36)$.
- $n = -77/2$ or $n = 36$.
- **Solution...**
- Since n can **not** be negative...
- Number of terms = $n = 36$.

12. Solution:

- a = First term = 10.
- d = Common difference = 6.
- S_n = Sum = $1896 = (n \div 2)(2a + d(n - 1))$.
- **Solving the equation...**
- $1896 = (n \div 2) \times (2 \times 10 + 6 \times (n - 1))$
- $1896 = (n \div 2) \times (20 + 6n - 6)$.
- $1896 = (n \div 2) \times (6n + 14)$.
- $3792 = n \times (6n + 14)$.
- $3792 = 6n^2 + 14n$.
- $0 = 6n^2 + 14n - 3792$. ← Note: **Quadratic equation**! Solve using the quadratic formula or by factorizing.
- First simplify by dividing both sides by 2...
- $0 = 3n^2 + 7n - 1896$.
- Then factorize...
- $0 = (3n + 79)(n - 24)$.
- $n = -79/3$ or $n = 24$.
- **Solution...**
- Since n can **not** be negative...
- Number of terms = $n = 24$.

13. Solution:

- a = First term = 5.
- d = Common difference = 7.
- S_n = Sum = $1727 = (n \div 2)(2a + d(n - 1))$.
- **Solving the equation...**
- $1727 = (n \div 2) \times (2 \times 5 + 7 \times (n - 1))$

- $1727 = (n \div 2) \times (10 + 7n - 7)$.
- $1727 = (n \div 2) \times (7n + 3)$.
- $3454 = n \times (7n + 3)$.
- $3454 = 7n^2 + 3n$.
- $0 = 7n^2 + 3n - 3454$.   ⬅ Note: **Quadratic equation**! Solve using the quadratic formula or by factorizing.
- $0 = (7n + 157)(n - 22)$.
- $n = -157/7$ or $n = 22$.
- **Solution...**
- Since n can **not** be negative...
- Number of terms = $n = 22$.

14. Solution:

- a = First term = -8.
- d = Common difference = 3.
- S_n = Sum = $161 = (n \div 2)(2a + d(n - 1))$.
- **Solving the equation...**
- $161 = (n \div 2) \times (2 \times -8 + 3 \times (n - 1))$
- $161 = (n \div 2) \times (-16 + 3n - 3)$.
- $161 = (n \div 2) \times (3n - 19)$.
- $322 = n \times (3n - 19)$.
- $322 = 3n^2 - 19n$.
- $0 = 3n^2 - 19n - 322$.   ⬅ Note: **Quadratic equation**! Solve using the quadratic formula or by factorizing.
- $0 = (3n + 23)(n - 14)$.
- $n = -23/3$ or $n = 14$.
- **Solution...**
- Since n can **not** be negative...
- Number of terms = $n = 14$.

15. Solution:

- a = First term = 10.
- d = Common difference = 11.
- S_n = Sum = $388 = (n \div 2)(2a + d(n - 1))$.
- **Solving the equation...**
- $388 = (n \div 2) \times (2 \times 10 + 11 \times (n - 1))$
- $388 = (n \div 2) \times (20 + 11n - 11)$.
- $388 = (n \div 2) \times (11n + 9)$.

- $776 = n \times (11n + 9)$.
- $776 = 11n^2 + 9n$.
- $0 = 11n^2 + 9n - 776$.   ← Note: **Quadratic equation**! Solve using the quadratic formula or by factorizing.
- $0 = (11n + 97)(n - 8)$.
- $n = -97/11$ or $n = 8$.
- **Solution…**
- Since n can **not** be negative…
- Number of terms $= n = 8$.

16. Solution:

- n = Number of terms = 14.
- S_{20} = Sum = $(n \div 2)(2a + d(n - 1)) = (14 \div 2) \times (2a + d(14 - 1)) = 7 \times (2a + 13d) = 406$.
- Term(14) = $a + (14 - 1)d = a + 13d = 55$.
- **We have simultaneous equations…**
- Equation 1: $7 \times (2a + 13d) = 406$.
- Equation 2: $a + 13d = 55$.
- **Now solve the simultaneous equations…**
- Rearrange Equation 1:
- $7 \times (2a + 13d) = 406$.
- $2a + 13d = 406 \div 7$.
- $2a + 13d = 58$.
- $a + a + 13d = 58$.
- Since we know from Equation 2 that $a + 13d = 55$, we can substitute 55 for $a + 13d$ in the above:
- $a + 55 = 58$.
- $a = 58 - 55$.
- $a = 3$.
- Substitute the value of a back into Equation 2:
- $3 + 13d = 55$.
- $13d = 55 - 3$.
- $13d = 52$.
- $d = 52 \div 13$.
- $d = 4$.
- **Answer:**
- First term $= a = 3$.
- Common difference $= d = 4$.

17. Solution:

- $S_{12} = (n \div 2)(2a + d(n-1)) = (10 \div 2) \times (2a + d \times (10-1)) = 6(2a + 11d) = 246$.
- $S_{24} = (n \div 2)(2a + d(n-1)) = (20 \div 2) \times (2a + d \times (20-1)) = 12(2a + 23d) = 924$.
- **We have simultaneous equations...**
- Equation 1: $6 \times (2a + 11d) = 246$.
- Equation 2: $12 \times (2a + 23d) = 924$.
- We can simplify Equation 1:
- $2a + 11d = 246 \div 6$.
- $2a + 11d = 41$. ← Let's call this Equation 3.
- We can simplify Equation 2:
- $2a + 23d = 924 \div 12$.
- $2a + 23d = 77$. ← Let's call this Equation 4.
- **Solve the simultaneous equations...**
- First combine the equations by doing Equation 4 - Equation 1:
- $(2a + 23d) - (2a + 11d) = 77 - 41$.
- $12d = 36$.
- $d = 36 \div 12$.
- $d = 3$.
- Substituting back into Equation 3...
- $2a + 11 \times 3 = 41$.
- $2a + 33 = 41$.
- $2a = 41 - 33$.
- $2a = 8$.
- $a = 8 \div 2$.
- $a = 4$.
- **Solution...**
- First term = $a = 4$.
- Common difference = $d = 3$.

18. Solution:

- d = Common Difference = 7.
- Last term = $a + d(n-1) = 148$.
- Arithmetic mean = $\frac{1}{2}(2a + d(n-1)) = 81.5$.
- **We have simultaneous equations...**
- Equation 1: $d = 7$.
- Equation 2: $a + d(n-1) = 148$.
- Equation 3: $\frac{1}{2}(2a + d(n-1)) = 81.5$.
- We can simplify Equation 3:
- $2a + d(n-1) = 81.5 \times 2$.
- $2a + d(n-1) = 163$.
- $a + a + d(n-1) = 163$.

- Since we know from Equation 2 that $a + d(n - 1) = 148$, we can substitute 148 for $a + d(n - 1)$ in the above, and solve for a:
- $a + 148 = 163$.
- $a = 163 - 148$.
- $a = 15$.
- We can substitute the values of a and d into Equation 2 and solve for n:
- $15 + 7(n - 1) = 148$.
- $15 + 7n - 7 = 148$.
- $8 + 7n = 148$.
- $7n = 148 - 8$.
- $7n = 140$.
- $n = 140 \div 7$.
- $n = 20$.
- **Solution…**
- First term $= a = 15$.
- Number of terms $= n = 20$.

19. Solution:

- $n =$ Number of terms $= 24$.
- Last term $= a + d(n - 1) = -99$.
- Arithmetic mean $= \frac{1}{2}(2a + d(n - 1)) = -41.5$.
- **We have simultaneous equations…**
- Equation 1: $n = 24$.
- Equation 2: $a + d(n - 1) = -99$.
- Equation 3: $\frac{1}{2}(2a + d(n - 1)) = -41.5$.
- We can simplify Equation 3:
- $2a + d(n - 1) = -41.5 \times 2$.
- $2a + d(n - 1) = -83$.
- $a + a + d(n - 1) = -83$.
- Since we know from Equation 2 that $a + d(n - 1) = -99$, we can substitute -99 for $a + d(n - 1)$ in the above, and solve for a:
- $a - 99 = -83$.
- $a = -83 + 99$.
- $a = 16$.
- We can substitute the values of a and n into Equation 2 and solve for d:
- $16 + d(24 - 1) = -99$.
- $16 + 23d = -99$.
- $23d = -99 - 16$.
- $23d = -115$.
- $d = -115 \div 23$.

- $d = -5$.
- **Solution…**
- First term = $a = 16$.
- Common difference = $d = -5$.

20. Solution:

- n = Number of terms = 100.
- S_{100} = Sum = $(n \div 2)(2a + d(n-1)) = (100 \div 2) \times (2a + d(100-1)) = 50(2a + 99d) = 2150$.
- Term(100) = $a + (100 - 1)d = a + 99d = 41.3$.
- **We have simultaneous equations…**
- Equation 1: $50(2a + 99d) = 2150$.
- Equation 2: $a + 99d = 41.3$.
- **Now solve the simultaneous equations…**
- Rearrange Equation 1:
- $50(2a + 99d) = 2150$.
- $2a + 99d = 2150 \div 50$.
- $2a + 99d = 43$.
- $a + a + 99d = 43$.
- Since we know from Equation 2 that $a + 99d = 41.3$, we can substitute 41.3 for $a + 99d$ in the above:
- $a + 41.3 = 43$.
- $a = 43 - 41.3$.
- $a = 1.7$.
- Substitute the value of a back into Equation 2:
- $1.7 + 99d = 41.3$.
- $99d = 41.3 - 1.7$.
- $99d = 39.6$.
- $d = 39.6 \div 99$.
- $d = 0.4$.
- **Answer:**
- First term = $a = 1.7$.
- Common difference = $d = 0.4$.

Conclusion

Well done for getting to the end. I hope you enjoyed this book!

For more about arithmetic sequences, and related topics, please go to:
http://www.suniltanna.com/arithmetic

If you enjoyed this book or it helped you, please post a positive review on Amazon!

To find out about other educational books that I have written, please go to:

- For math books: http://www.suniltanna.com/math
- For science books: http://www.suniltanna.com/science

And remember: If you enjoyed this book or it helped you, please post a positive review on Amazon!

Printed in Great Britain
by Amazon

37306432R00044